The Backyard
Railroader

The Backyard Railroader

Building and Operating a Miniature Steam Locomotive

Jeff Frost

McFarland & Company, Inc., Publishers
Jefferson, North Carolina

All photographs and drawings are by the author
unless otherwise indicated.

Library of Congress Cataloguing-in-Publication Data

Names: Frost, Jeff, author.
Title: The backyard railroader : building and operating
a miniature steam locomotive / Jeff Frost.
Description: Jefferson, North Carolina : McFarland & Company, Inc.,
Publishers, 2019 | Includes index.
Identifiers: LCCN 2018052800 | ISBN 9781476672816
(softcover : acid free paper) ∞
Subjects: LCSH: Locomotives—Models—Handbooks, manuals, etc. |
Steam locomotives—Models—Handbooks, manuals, etc.
Classification: LCC TJ630 .F76 2019 | DDC 625.1/961—dc23
LC record available at https://lccn.loc.gov/2018052800

British Library cataloguing data are available

ISBN (print) 978-1-4766-7281-6
ISBN (ebook) 978-1-4766-3164-6

Front cover inset the author's father bought this little Shay locomotive in 1989 (photograph by L. Lee Frost); foreground author Jeff Frost is running the 4-6-0 in Pennsylvania while his sons are enjoying their ride (photograph courtesy of the author)

Printed in the United States of America

McFarland & Company, Inc., Publishers
Box 611, Jefferson, North Carolina 28640
www.mcfarlandpub.com

Table of Contents

Preface

This book is not meant only for beginners or for people who are simply interested in the live steam hobby or miniature steam locomotives. It is also for those who have been in the hobby for a while. In this book are different problems my father (Donald S. Frost) and I experienced during our many years in the hobby as well as questions encountered by other hobbyists. It is meant to be used as a resource to help beginners to get started, and for those who are more experienced, to help troubleshoot particular problems and to improve the performance of miniature steam locomotives following practices used in full-size trains. This includes explanations of why things are done in a certain way. While most of the material covers specific aspects of steam locomotives (which can be applied to locomotives of pretty much any size), some of the material can apply to "diesels" or cars or steam traction engines. The basics of laying track, wheels and axles are covered, along with my experiences in the hobby.

Steam locomotives, while inanimate objects, can seem to be alive. Each engine has its own personality and temperament, and as the engine is running a perceptive engineer can hear it speaking. Noise from the stack (exhaust from the cylinders), noise from the fire, noise of the rods, the appearance of the smoke, and the steam pressure all talk to the engineer. The engineer can tell how the engine is doing and if there is anything the engine needs. That is part of the fun of the hobby.

Unlike an automobile, a steam locomotive requires quite a bit of attention and tweaking. Even if all the parts are built exactly to plan, errors of one kind or another will inevitably creep in and some adjustments will be required. However, a steam engine will run even when some things are quite a bit off. There is an old saying that a steam engine can give you 90 percent of its capabilities when it is only performing at 50 percent, but a diesel will give you 0 percent if it is performing at 95 percent. On the other hand, a steam locomotive by its nature is starting to destroy itself and wear itself out the minute it starts running.

I was born and raised in Traverse City, Michigan. My dad got started in the live steam hobby in 1989 (when I was around six years old) and between 1990 and 2001, he and I built three 7.5" gauge steam engines. I truly grew up in the hobby.

When I was little I remember watching one of Walt Disney's cartoons, titled "Out of Scale." Donald Duck was the engineer on a miniature steam locomotive (a 4-4-0; Walt Disney had built a 7.25" gauge 4-4-0 in the 1940s). Donald Duck had a miniature railroad in his backyard, complete with miniature buildings, trees, streets and farms. Unsurprisingly, I wanted to have one! Dad was also interested in trains and had built an HO layout and still had his old Lionel set from when he was a boy.

On days when Mom was working and Dad was not, Dad would take my sister and me on different outings, typically looking for something train-related. Around 1988 Dad had heard of a man (we later discovered this was Bill Fitt) living north of Cadillac, Michigan (a little less than an hour south of where we lived), who had a large-scale, outdoor riding railroad. The first couple of times we went out looking for it we didn't find it. When Dad finally stopped at a local store and asked the store owner if he knew about this railroad, we got lucky and the owner gave us directions. Bill Fitt was more than willing to show us his small shop and engines and he also encouraged Dad to get into the hobby. Dad, who is now a retired doctor, didn't know much about building or running a steam engine or machining parts, but Bill encouraged him by saying that if you have the time, money and willingness to learn and to make mistakes (resulting in so much scrap) that anyone could do it.

Dad bought his first steam locomotive, a small 7.5" gauge Shay, in 1989. He soon started buying tools and machines to set up a shop. Around 1990 he started to make his first engine. As I got older and started learning how to machine parts, I got more involved. By May 2001, Dad and I had finished our third steam locomotive. Starting in the summer of 2004, Dad and I worked on and ran a 15" gauge steam locomotive, a 4-4-2, that ran around Clinch Park (from 1982 until 2011) in Traverse City. During my college breaks, we would also do maintenance to the train. After I graduated from Michigan Technological University, I started working at the Strasburg Rail Road (a steam tourist railroad in southeastern Pennsylvania) as a machinist and mechanic. In 2008 I started in engine and train service and in 2010 I became a conductor and engineer and am also a locomotive inspector as well as an amusement ride inspector.

I live in Lancaster, Pennsylvania, with my two sons. In addition to my interest in trains, steam engines and the live steam hobby, I am also interested in music. My main instrument is the trombone and I play locally with the Elite Brass. I am also a composer.

This book got its start when I began posting my writings online. In response, people started asking me questions about track, designs, problems and what might be causing

***Left:* Jeff is an engineer and runs and fires the engines at Strasburg Rail Road. Here he is at the throttle of #7312, a former Canadian National 0-6-0, around 2008. *Right:* Jeff is also a trombonist and composer and arranger.**

them and asking me how to solve them. Even some newcomers to the hobby were asking about some of the more basic things. People encouraged me to write a book about the hobby so I started saving my responses.

I enjoy learning about steam locomotives and passing along the knowledge to others. When Dad started in the hobby, there really wasn't much readily available material that pertained to the live steam hobby or explaining things to do with it. There were books that touched on a few topics and maybe you could find some old steam books at the local library. Now, there is some limited material on the internet. Around 2010 I started writing about the things that Dad and I have learned (some by reading books and some learned the hard way) in order to pass the knowledge along to others. There are multiple aspects to the hobby that people can enjoy (running a train, building scale buildings, building cars, building "diesels," building steam locomotives, laying track, tweaking or improving existing engines, highly detailing engines and cars, etc.). Dad and I mainly enjoy building trains and tweaking or improving the steam engines. While the engines we built are not super detailed, they run very well.

Abbreviations

0-4-0, 2-4-0, etc. Indication of locomotive wheel arrangement using Whyte notation. The first number indicates the number of unpowered leading wheels, the middle number indicates the number of driven wheels, the last number indicates the number of trailing wheels.

CFR—Code of Federal Regulations. The volumes of codified rules enacted and published by agencies of the executive branch of the U.S. government, including those governing the design, construction and operation of boilers.

FRA—Federal Railroad Administration.

IBLS—The International Brotherhood of Live Steamers. Founded in 1932, the organization works to connect live steam hobbyists through meets.

ID—inner diameter. The diameter of the inside of a tube or any circular part.

MAWP—maximum allowable working pressure.

MTP—model taper pipe thread. Fine tapered pipe thread standard used in model engineering.

NPT—national pipe thread. American standard tapered pipe thread.

OD—outer diameter. The diameter of the outside of a circular part.

TDS—total dissolved solids.

UT—ultrasonic testing.

Safety Is Rule No. 1

Safety is very important in this hobby. Always remember that "Safety Is Rule No. 1" or "In Case of Doubt, Choose the Safest Course." But what does this look like? If one is unsure what the next step should be (machining, operating a steam or diesel engine, moving heavy equipment, setting up a machine shop), the safest course would be to stop and ask someone for advice or help and/or read books on the topic to further one's knowledge and understanding. Safety is something that is done intentionally; it does not happen by chance. Safety has different meanings and actions depending on what is happening.

Basic Machine Shop Safety

When doing any machining or sawing, safety glasses or a clear face shield must be used to protect the eyes from debris.

Any piece of loose clothing, gloves, long hair, rings, jewelry, etc., can get caught in the machine. The machine won't know the difference between metal or wood and flesh, and the machine can pull you into it. Do not stop chucks/work from spinning by using one's hands. Do not put a finger inside a rotating part.

Do not wear rings, jewelry or loose clothing when operating power machinery. Remove gloves and always wear safety glasses when operating power machinery. Keep long hair tied up to keep it from getting caught in the machinery, and keep oily rags in properly marked containers (suitable for oily rags) so they don't catch fire.

Clean up any oil or grease slicks on the floor and sweep up chips and other debris from the floor to minimize slipping hazards.

Basic Welding/Torching Safety

Keep the welding/torch cutting area clear of flammable items. When oxy-acetylene welding or cutting, wear heat-resistant gloves and a face shield or goggles that have a minimum shade color of 5. *Never* have the acetylene pressure too near 15 psi as acetylene becomes unstable. Always keep the oxygen and acetylene bottles chained down and upright (in a cart, to a solid post) to keep them from falling over and breaking off the valve on top of the bottle. Whenever transporting bottles, always put the protective cover over the valve to protect the valve from damage or being accidently sheared off. *Never* lubricate the regulator valves with oil; in regular atmosphere (21 percent oxygen) oil will burn, in 100 percent oxygen it could explode.

When using an arc, mig or tig welder, wear heat resistant gloves, a face shield that has a minimum shade color of 10 and welder's leathers or clothes to protect any bare skin from UV light. *Never* look at the arc with the naked eye when welding; this can burn the retina.

Finally, it is a good practice to keep a fire extinguisher near the welding/cutting area.

Basic Safety with the Track and Riding

The track should be kept away from pinch points such as trees or buildings or other structures (in other words, provide ample clearance), so if a derailment occurs, people aren't stuck between the train and an immovable object. If a railing is applied to a bridge, there should be enough room between the train and railing to be able to walk and not get pinched.

It is a good idea to keep grades to 2.5 percent or less if possible. While climbing a hill can be a mechanical problem, the bigger problem is not being able to stop when running down a hill. Curves on the downhill side can help slow the train down, as they add resistance.

On curves, the outside rail should be slightly higher than the inside rail (superelevation). If the rails in a curve are level, it won't feel quite right and people may lean, causing problems with balance. If the outside rail is lower than the inside rail, it really won't feel right and cars may tip over. On straight track, the rails should be level with each other.

When figuring out what radius curve will be used, decide what equipment/engines will operate on the track. Large 1:8 scale steam engines need a 40'–60' radius curve, whereas smaller steam engines and medium size diesels need 20'–40' radius curves. Laying the largest possible radius curve is a good rule of thumb.

The track is not meant to be walked on. The rails are tripping hazards and walking on the track can kick stones up onto the track, potentially causing a train to derail.

When riding on miniature trains, riders should sit still and not rock the car or move about when the train is moving. This includes leaning to one side. A human is much larger proportionately than the miniature train (someone who weighs 100 pounds is proportionally 51,200 pounds in 1:8 scale) and rocking, leaning or moving about in the car can cause the car to tip over (considering the high center of gravity with the person in the car) or derail. If there is only one rider on a car, the best practice is to ride in the center or towards the front of the car. If there is too much weight in the back of the car, the front truck might not have enough weight on it and could derail easily.

Basic Safety with Running Trains

Keep the speed safe. When sitting close to the ground, 5 mph (7.3 ft/min) can seem quite fast. The miniature trains don't stop quickly and if a car derails at a high speed, more damage can be done to the track, cars and riders. Most miniature engines could do 20 mph if let loose.

With a steam engine, keep the water in the water sight glass and keep the sight glass in good working order to prevent overheating and damaging the crownsheet/boiler.

If cars have couplers and safety chains, use the safety chains always. The safety chains will keep the cars from rolling downhill if the couplers break or become uncoupled.

Always be looking forward or in the direction the train is running. Since the trains don't stop quickly, be alert for obstacles on the track, switches that are set wrong or other trains so you can stop easily before derailing or hitting anything.

When working on an engine that is on a stand up off of the ground, make sure there are clamps on the stand to prevent the engine from rolling off. Ensure that the stand is of a good design that will easily support the weight of the engine. Small engines may weigh 200 pounds, while larger engines may weigh upwards of 700 pounds.

These miniature engines can pull a lot of weight. That means starting and stopping will take time and distance. An engine that weighs around 400 pounds could pull a 2,500-pound train on a 2 percent grade. That is a lot of weight to start and stop.

Safety Summary

Machinery such as lathes, mills, drill presses can be dangerous. Scale railroad equipment such as steam engines (a medium size 1:8 scale engine can weigh around 500 pounds) and cars (1.5" scale cars can weigh 75–150 pounds, sometimes more) can be unforgiving. All need to be treated with respect and it is important that the person operating them needs to be knowledgeable about the equipment they are using.

Getting Started

Welcome to the Live Steam hobby. The hobby is full of people with varying backgrounds (I have met hobbyists who are doctors, musicians, teachers, lawyers, clergymen, machinists, mechanics, railroaders, and many more) who share the love of scale railroading. Not everyone who enters the hobby knows how to read blueprints or how to machine parts. Many people learn how to do these things by reading books, being mentored by others and learning by doing. As long as people have the determination, time to learn and willingness to learn from mistakes, they can build their first engine/car/track. The hobby can be personally rewarding. There are multiple aspects of the hobby to enjoy: building engines and cars, machining, painting, improving existing equipment, detailing equipment, running and pulling people, running and setting out and picking up freight cars in an "operations" style meet, adding track side structures such as house and factories, and much, much more.

When first getting into the hobby it is a good idea to find and join a local club. The members are a great source of knowledge and experience and they can help to teach how to do things and to recommend ideas, engines, machinery and books. The most common scales are 1" (1:12) and 1.5" (1:8) but the gauges for 1" scale are 4¾" or 5" and for 1.5" scale they are 7¼" (in the eastern part of the U.S. and in Europe, New Zealand and Australia) and 7½". It is a good idea to build equipment that can be run on a local club track instead of starting with a size that is not used in the area and thus needing to build a track for the equipment to run on. While people certainly have built sizes that are not common, it is much easier for new people in the hobby to have equipment that can run on a local track.

To Buy or to Build?

When I am asked if someone (who I generally know nothing about) should buy an engine or build one, my general response is that if they want to build an engine, they should build one (there are casting kits available). But since it can take years to build a steam engine, I highly recommend buying an engine to run in the meantime. Building an engine, once the machinery has been bought, is generally less expensive than buying an engine that is ready to run. Building an engine is time consuming, but is a good learning experience (reading prints, machining parts accurately, learning how the machine works, etc.) and it also helps a person learn the skills to maintain the engine. Someone who buys an engine that is running will have to learn how to maintain it as things wear out. Either way there is a lot of learning. For a person who does not do any machining, I would strongly recommend buying some machinery and start practicing (i.e., making scrap) and teach yourself (with the help of books as well as other machinists and live steamers). The way to learn is to do it and learn from mistakes.

When my father (generally referred by me as Dad in the rest of this book) first got into the hobby, he bought a small two-cylinder Shay that had been built by Dutch Boysen of Yakima, Washington. The Shay had apparently sat around for six or so years due to deterioration in Dutch's health and was not the greatest runner or steamer as not all the bugs had been worked out of it. Because Dad was new in the hobby, he chalked most of the problems up to just being inexperienced. Later on I met people who knew Dutch and they said that his other two engines were great engines, but he only ran the Shay for maybe a year tops before he had to get out of the hobby. While the Shay was not the greatest engine, it gave us something to run while we built the 0-4-0 (later modified into a 2-4-0).

For a beginner interested in more modern engines I would recommend Railroad Supply 0-4-0, 2-6-0/0-6-0, Little Engines 0-4-0, modern 2-6-0 for smaller (easier to move and handle) engines. For more old-time engines, I would go with Railroad Supply 4-4-0, Little Engines 4-4-0, old time 2-6-0 and Allen Models 2-6-0 and 4-6-0. These engines generally weigh 250–350 pounds and can easily be moved by one person. They can pull a decent load (a RRSC 0-4-0 can pull 1,520 pounds on a 3 percent grade with 52' radius curve, 2,400 pounds on a 2 percent grade with 75' radius curve or in theory around 65,000 pounds on flat, level track) and are good steaming and running engines.

Figure 1. Dad bought this little Shay locomotive in 1989 from Dutch Boysen (courtesy L. Lee Frost).

Figure 2. This is the first locomotive that Dad and I built. She started out as a Railroad Supply 0-4-0 and was first running in May 1996 and was modified to a 2-4-0 in the early 2000s.

For the first build, Dad decided on a RRSC 0-4-0 (later modified to a 2-4-0) as it would handle down to a 15' radius curve. At that time, Dad was looking at a small 250' loop in the backyard and it would have around a 20' radius curve. Dad started her in 1990 and she first ran in 1996. The new larger tender and lead truck were added around 2000–2002.

Next we built a RRSC 2-8-2. We started her in the late fall of 1996 and she was running in May 1999. She is a great work horse, but at 900 pounds for the engine, she was too big for Dad to run and move by himself.

Figure 3. This is the second locomotive that Dad and I built. She is a Railroad Supply 2-8-2 USRA heavy Mikado and was first running in May 1999.

Figure 4. This is the third locomotive that Dad and I built. She is a modified Allen Models 4-6-0 and was first running in May 2001 (courtesy Donald S. Frost).

So Dad decided on an Allen Models 4-6-0, which weighs around 350 pounds and was a good balance between the 2-4-0 and 2-8-2. We started the 4-6-0 in September 2000 and she first ran in May 2001. As part of the building, I designed and fabricated the Walschaert valve gear for her.

Building and maintaining a steam engine can be very fulfilling, but it takes time and commitment.

Buying a Steam Engine

Since external appearance is not always a sign of the quality of a running engine (just because an engine is highly detailed doesn't mean it runs well; just because an engine has minimum detail doesn't mean it is a poor runner), a buyer needs to have more objective things to consider. Here is a basic list of what I would look for when buying a steam engine. These are in no particular order. While this is written more for steam engines, items in this list also can apply to diesels and cars.

If the engine is not under steam or running:

1. Do the flanges have a nice rounded contour and is there a decent radius where the flange meets the tapered tread? Does the tread have a taper, or is it worn flat? (Sharp, worn or square flanges [which can chew up the rail; thin flanges can break easily] need to be turned to a new and proper shape. A wheel that does not taper so it is larger at the flange to smaller on the outside of the wheel may still have use left but is close to the end of its life.) Do the wheels appear tight on the axles or is oil or grime working out between the axle and the bore of the wheel? This signifies that the wheel is loose on the axle.

2. Does the engine roll easily? Is there any binding in the rods or any tight or binding spots as the wheels rotate? The engine should roll easily without any grinding or metal-on-metal noises. There will typically be a smooth resistance from the pistons.

3. How loose or tight are the rods? The rods should be able to move side to side some so they don't bind on the crank pin caps as the axles shift in the driving boxes while going around curves. The rods should be able to move forward and backward slightly (I like to have .005" [about the thickness of five hairs] clearance between the crank pins and the side rod bushings) so they don't bind on the crank pins and don't wipe off the oil/grease lubricant. If they move excessively (.020"+) they should be replaced in the near future. Are there any cracks seen in the rods or valve gear?

4. How loose or tight is the valve gear? If the culminated slop at the valve (the valve can be pushed and pulled without moving the engine) is 0.040–0.055" ($^{3}/_{64}$"+), I rebuild the valve gear (mainly replacing bushings and pins) as that is enough to have a noticeable effect on the timing/valve movement.

5. When pushing the engine back and forth, can you hear air being pushed out of the cylinders at the stack or open cylinder cocks? I have heard of stories where engines were sold without pistons or valves. If air movement is not noticeable when pushing the engine, then one would start to question if the valves and pistons are in place.

6. Does the engine have a lubricator (mechanical or hydrostatic) to lubricate the piston and valves? If it is mechanical, does the pawl engage and rotate the ratchet as you move the engine? Do the piston and valve rods have a slight covering of oil, or are they rusted, showing minimal lubrication? If the cylinders have not been properly lubricated, they can be damaged or scored. Sometimes this requires new piston rings and reboring or honing the bores.

7. Does the boiler show any signs of leaks at welded joints or staybolts or signs of bulging on the outside of the boiler?

8. Look inside the firebox (use a light) through the firebox door and pay attention to the grates and burners. Do they look like they are really beat up, burned or damaged? Or do they look like they have been used but not abused?

9. While inspecting the firebox, look at the sheets and tubes inside the firebox. Is there any rust around the tubes, or on the sheets that can signify that the tubes or sheets are leaking? When was the boiler last given a hydrostatic test? A leaking boiler is an expensive fix.

10. Open up the smokebox and inspect the front tube sheet and tubes for signs of leaks and bulging. Inspect the steam lines as they exit the boiler and head to the cylinders for signs of leaks, decay or damage.

If you can look into the boiler (steam dome, plug openings), is the boiler decently clean of rust and scale, or is it filled with rust flakes and scale? A dirty boiler can be cleaned, but it takes time and the dirt may hide signs of decay and thin spots.

11. Does the tender roll easily? If you sit on the tender, do the springs support you or are they fully compressed? How does the inside of the water tank look? Is it fully rusted out?

12. Does the brake system appear to be connected with tie rods to operate all

brake shoes? If the brakes are manual, check to see if they work. If they are steam operated, check to see if the brake valve is piped up to a steam source (probably from the turret on the boiler) and piped to the brake cylinder.

13. Do the valves in the cab and any connected directly to the boiler function easily or are they seized up? Does the throttle open and close easily? Are there ways to get water into the boiler other than a handpump (which is very time consuming) such as an axle pump, injector, crosshead pump or steam driven pump?

14. Is the water sight glass (located in the cab, for the most part on the backhead) clean and do its valves operate easily? Is the bottom of the sight glass above the top of the crownsheet (the highest point in the firebox)? In some states and clubs, the bottom of the sight glass must be a minimum of ½" (0.500") above the top of the crownsheet. On my engines, I have a punched line on the backhead which marks the height of the crownsheet.

15. Can you rock the engine side to side easily and see the frame move in the driving boxes? Or is it stiff? If the springs are too stiff, the engine will ride hard which can beat it up over time. An engine can also be stiff if the driving boxes are stuck in the frame and don't move easily up and down in the frame.

If the engine is under steam, check these in addition to the above list:

16. Do the brakes work?

17. Does the chuffing of the engine sound even and square? Does the engine pulse when running? Does it sound good in forward and reverse?

18. If the engine is running in the corner (full forward or full reverse), start moving the Johnson bar closer to center/neutral. Does the engine still have four even exhaust, and sound good running in forward or reverse? A valve gear that is set up well will sound even in forward and reverse and the engine can be run close to center. (All three of the engines that Dad and I built run well just forward or just reverse of center.)

19. Is the sight glass operating well and can it be read easily?

20. Do the injectors, axle pump, crosshead pump or steam-driven water pump work well and does it keep up with the demand?

21. Are there any weird clacking noises or noises that sound like something is rubbing against the spokes of the wheels?

22. Is the lubricator using oil and lubricating the valves and pistons?

23. How well does the engine steam running light versus pulling a good-size train? The boiler should be able to keep up with the demand.

24. Does the engine ride well? Does the engine track well or does it derail easily? Does the engine take left and right curves and switches easily?

Casting Kit versus Building from Scratch

One common question I get after the "how much does an engine cost" question is, "How much money can be saved by designing and building an engine from scratch versus from a casting kit?" The answer is that you can save a lot of money, but be prepared to invest at least double the time to design and then fabricate the parts. For designing, a good working knowledge of the finished engine is needed in order to provide for holes and brackets for equalizing, brakes, valve gear, etc., in the frame while accessible. I would

Figure 5. The original drive wheels (new axles and crank pins) and new lead truck are installed under the new frame for the 4-6-4 (courtesy Donald S. Frost).

Figure 6. This is the front view of the new lead truck. The lead truck frame, journal boxes, springs and swing bolster can be seen (courtesy Donald S. Frost).

not recommend designing a steam engine from scratch for a first-time builder as it is a complicated process. It would be better to use a casting kit and design that works well.

For the 4-6-4 that Dad and I started "rebuilding" for a client several years ago, I have been doing the designing and Dad has been doing the building. The material for the frame cost around $500–$600, far less than what a typical casting kit for the frame

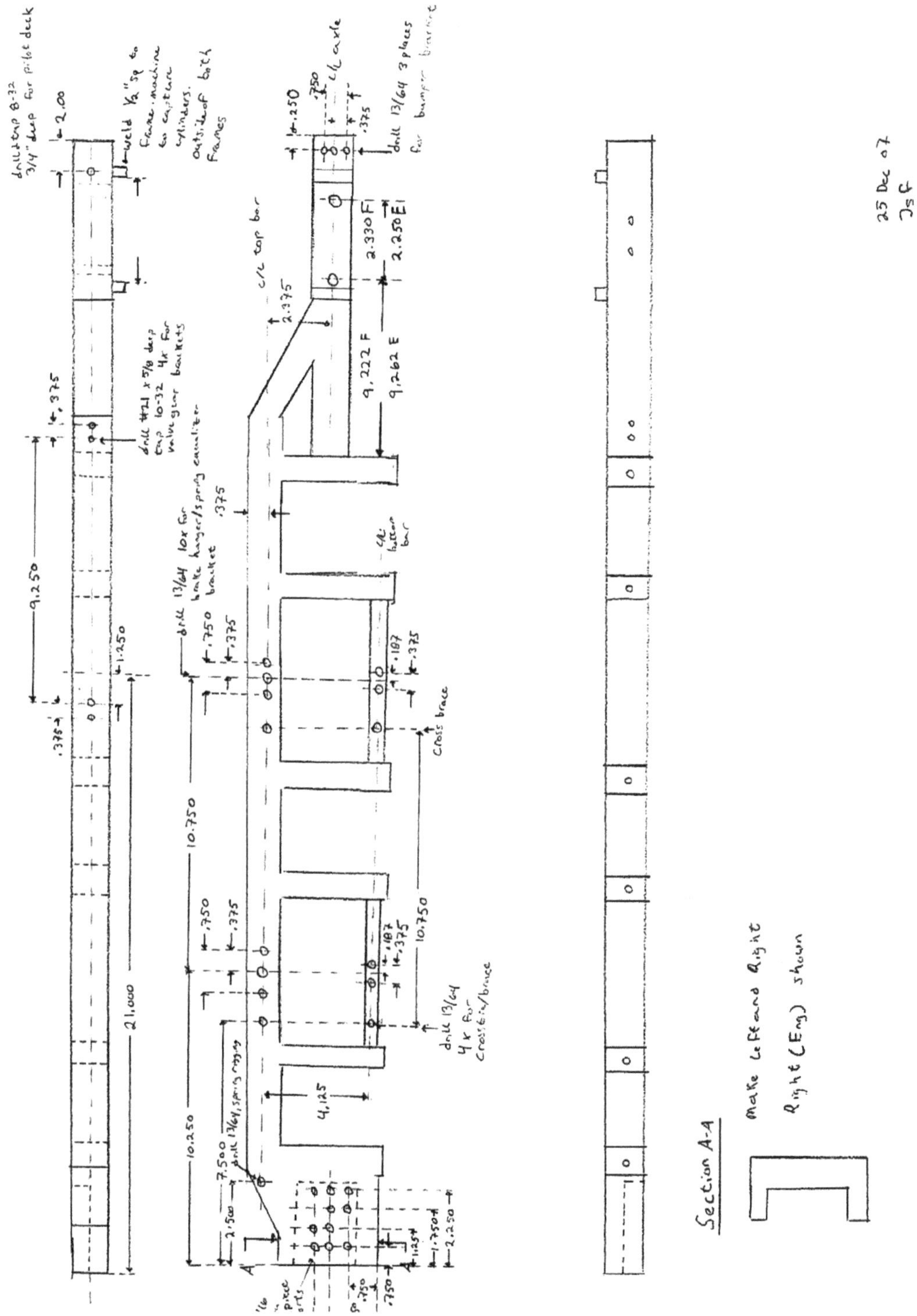

Drawing 1. This side frame drawing gives the hole placement on the side and top of the new frame for the 4-6-4. This includes holes for the frame spreaders, holes to attach the rear frame pieces, holes for the brake hanger brackets, cylinders, brake beam pillow blocks and the brackets that will be used for the valve gear and crosshead guide supports. This was made when we were still looking at using the original cylinder castings. It was later decided to fabricate a new cylinder block due to multiple issues with the castings.

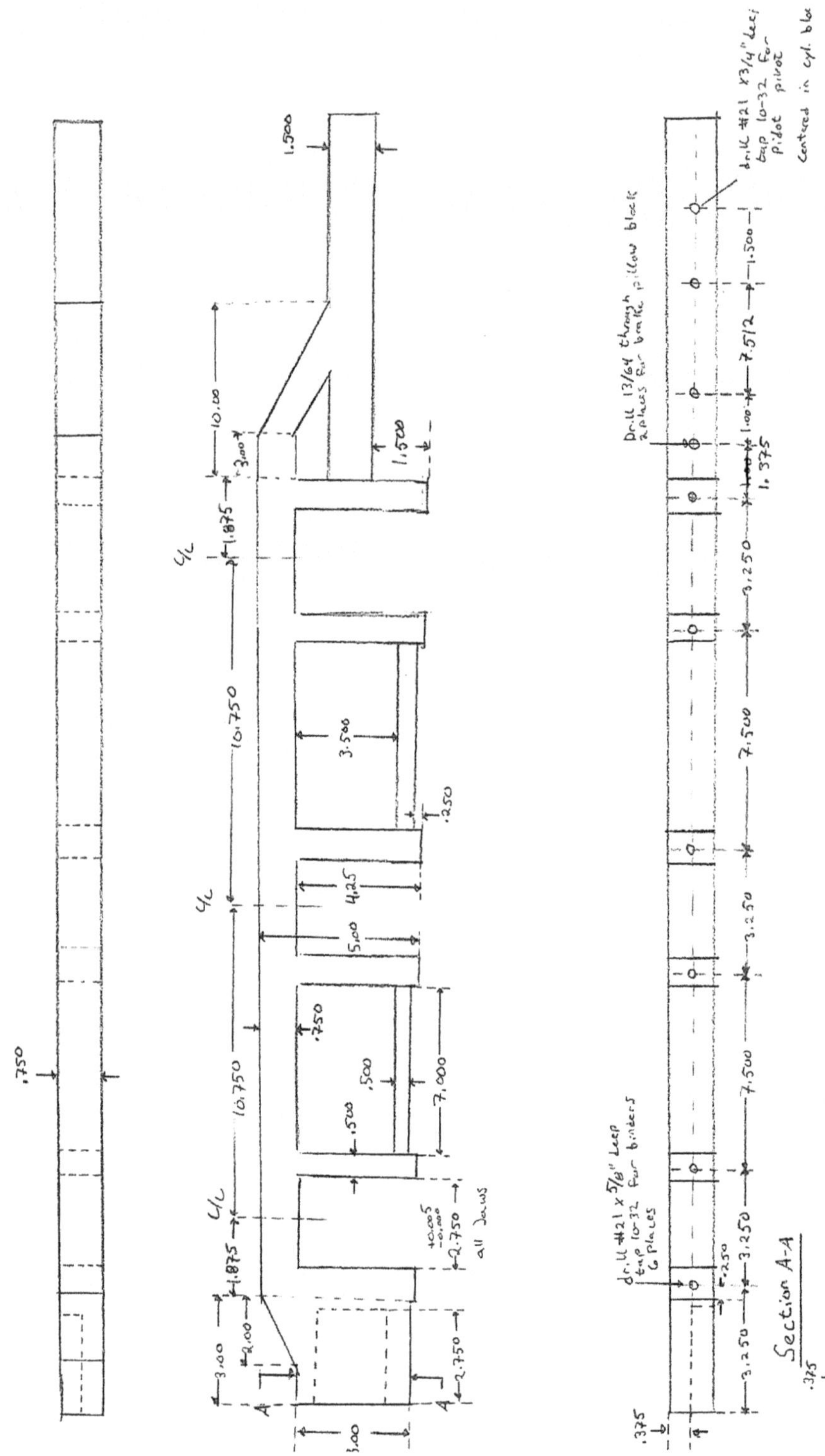

Drawing 2. This side frame drawing gives the machined dimensions of the openings for the driving boxes, axle centerlines, height differences between different frame members and the hole placement for the pedestal binders. The axle centerlines will be used when making the side rods, the dimensions of the openings of the pedestals will be used when making the driving boxes (made large enough to fit a ball bearing) and when making the pedestal binders.

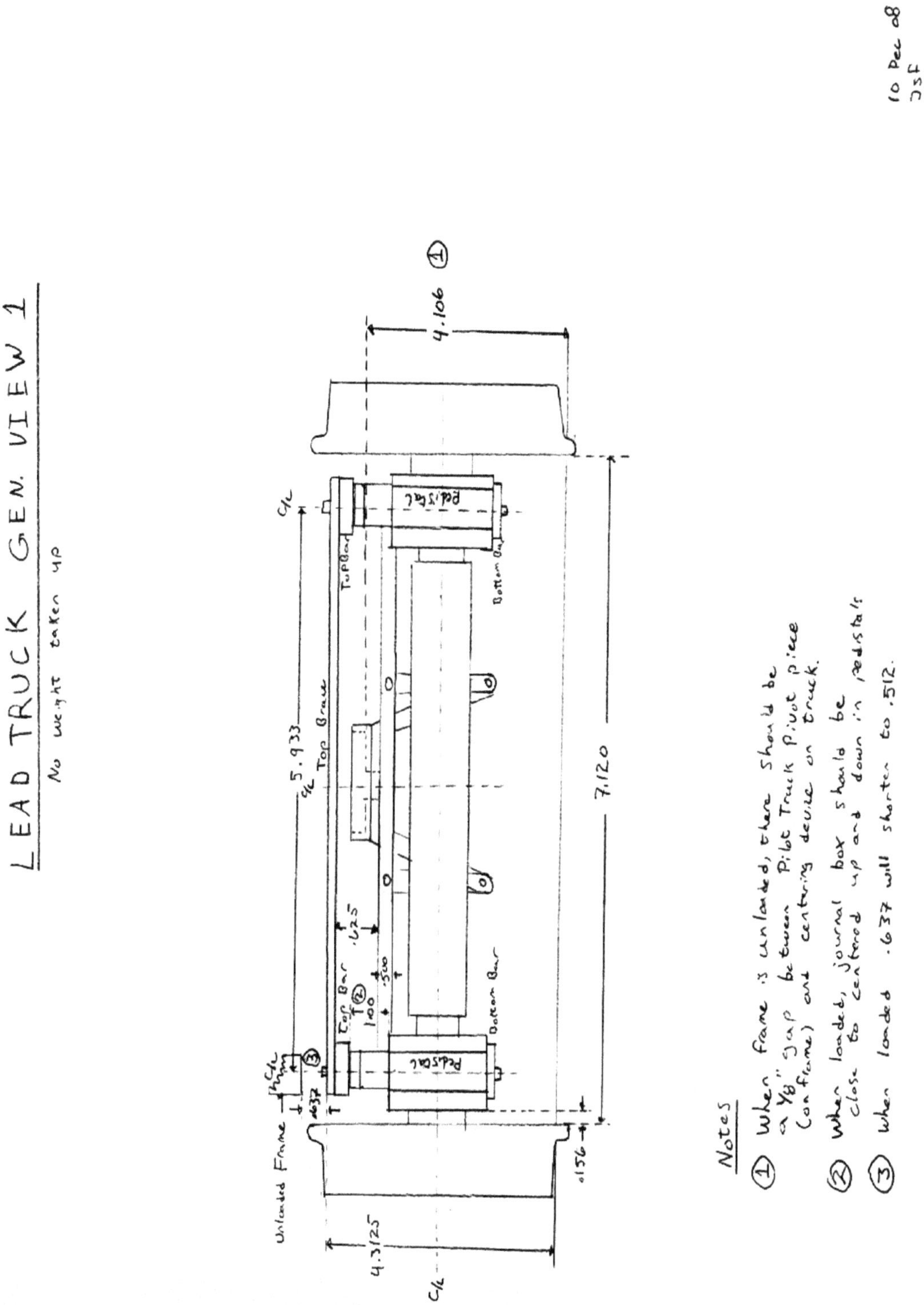

Drawing 3. The front view of the lead truck shows the placement of the frame and journal boxes in relationship to the wheel and the swing bolster and the proposed unload height of the bottom of the engine frame and the bottom of the swing bolster cup.

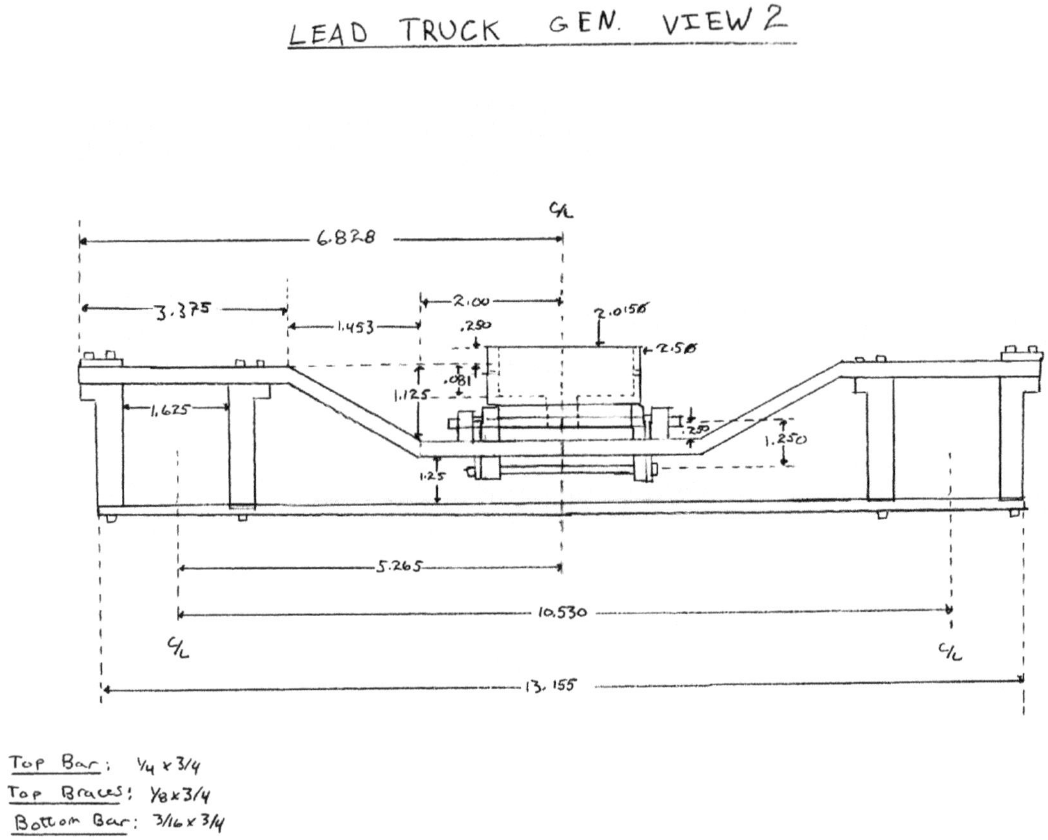

Drawing 4. The axle centerlines are spaced out to allow the cylinder block to fit between the lead truck wheels. The ball bearings for the axles were picked out and the pedestal jaw openings are large enough to fit a journal box that will house the bearings. The basic dimensions used to design the swing bolster are figured out in this drawing.

Figure 7. The side view of the lead truck shows the fit up of the swing bolster to the pivot on the frame. Note also that it is centered around the bolt holes for the cylinder block (courtesy Donald S. Frost).

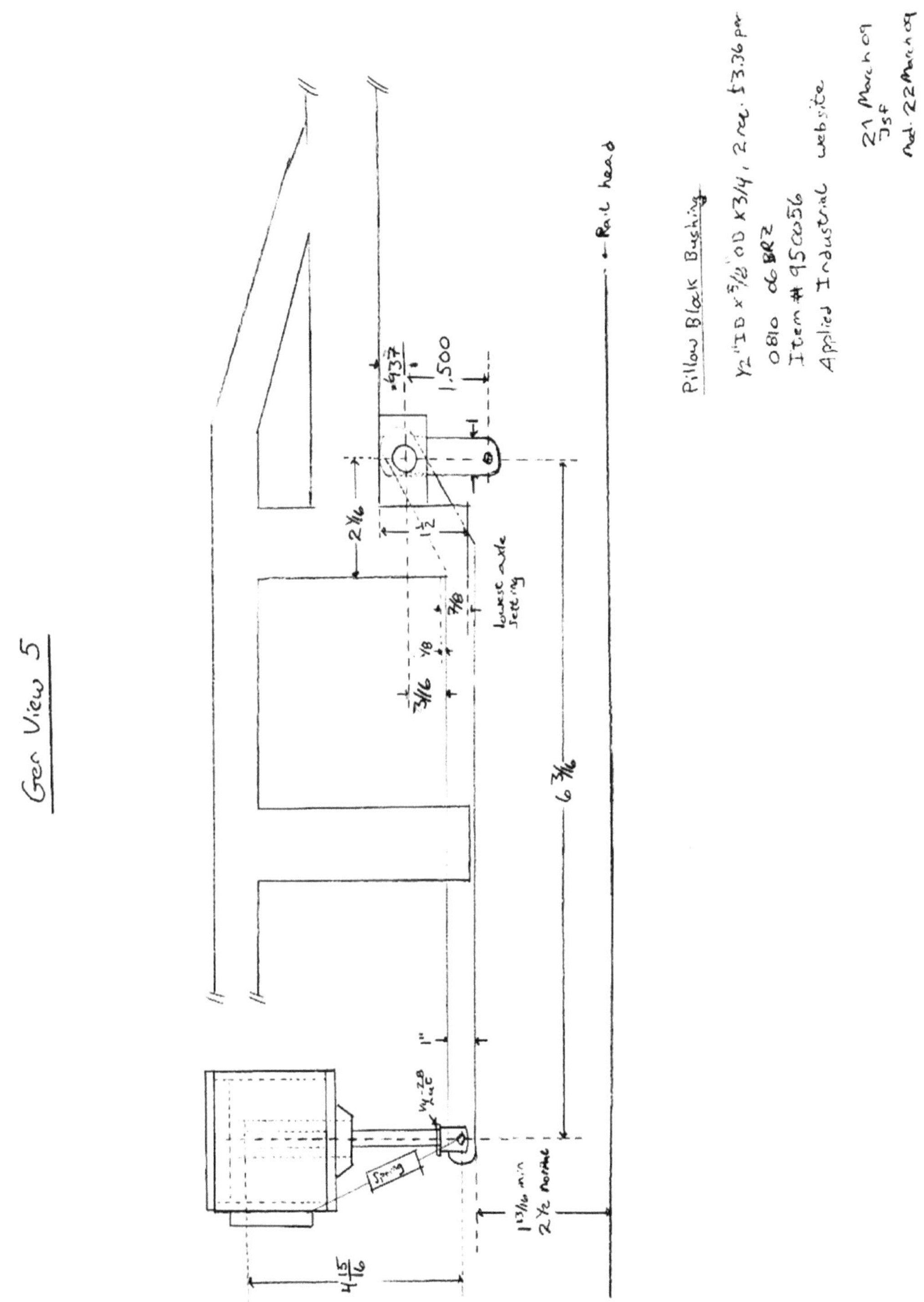

Drawing 5. When designing the brake lever that connects the brake cylinder and transmits the power to the brake rods the brake level needs to be able to fit to the engine frame and clear the lead driving axle. The brake cylinder will be connected to a bracket on the frame so the piston rod is in line with the hole in the end of the brake beam.

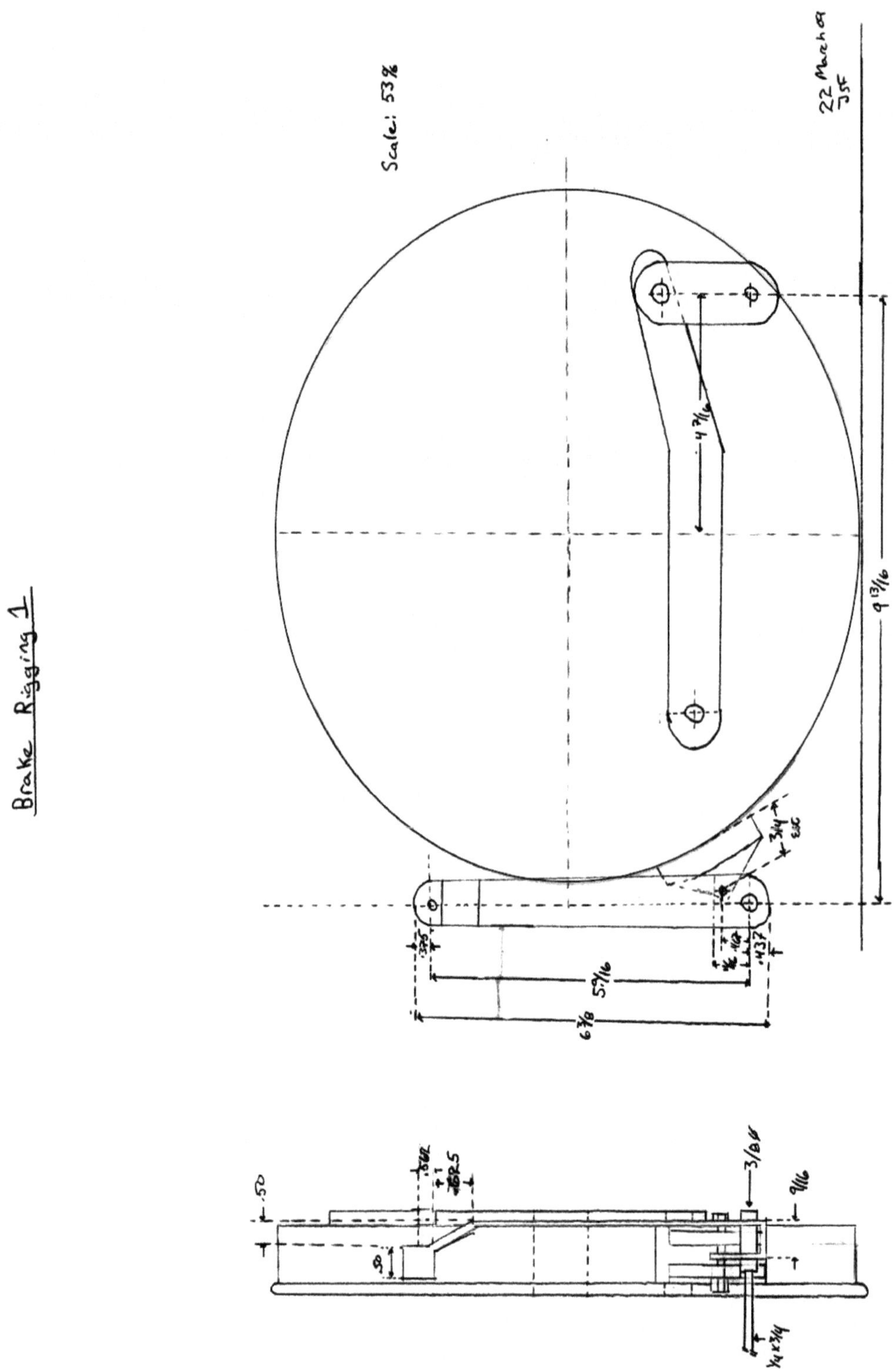

Drawing 6. The layout and dimensions of the brake hanger are determined to be able to make the brake hanger bracket and also the length of the tie rod to connect to the brake lever.

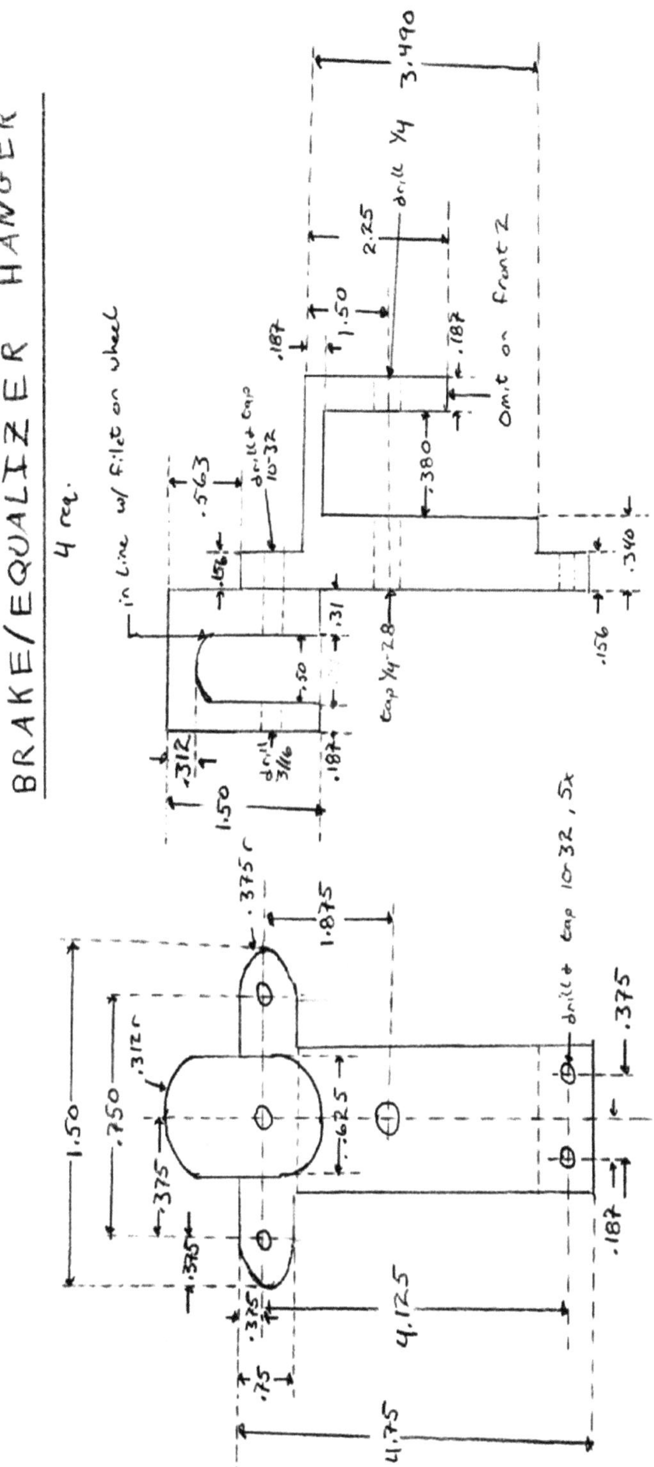

Drawing 7. Once the positioning of the brake hanger is determined, the brake hanger bracket can be designed to fit the holes that are already drilled in the frame and to correctly position the brake hanger in the proper alignment to the wheel. The back side of the bracket can also have a bracket to hold the equalizer, a part of the spring rigging/equalizing on the engine.

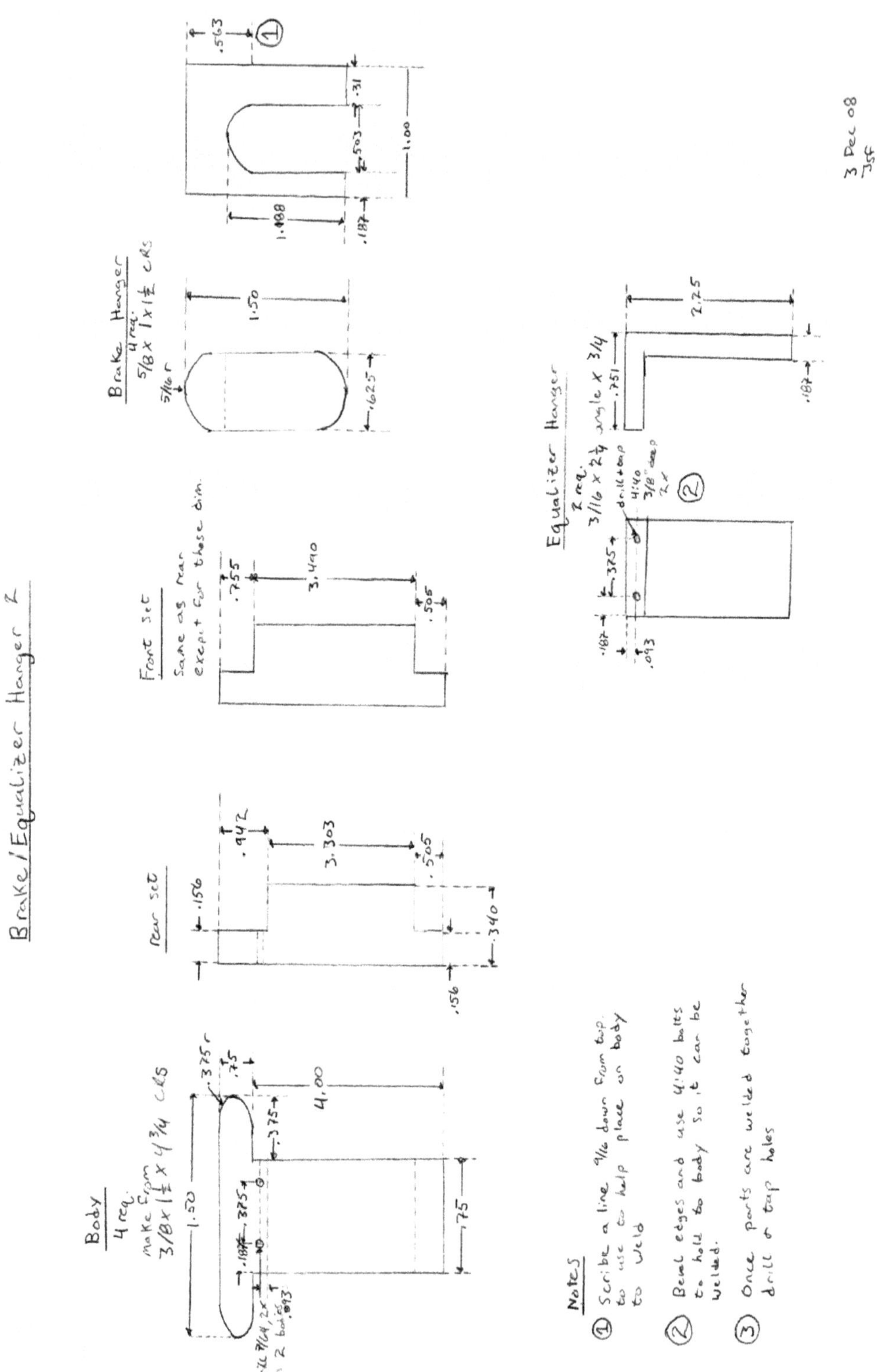

Drawing 8. The brake hanger bracket will be fabricated out of several pieces. This drawing shows the individual pieces to make the brake hanger bracket.

Top: Figure 8. The brake hanger bracket is installed on the frame between the drivers. The springs and spring rigging/equalizing have been designed, made and installed. The main side rod connecting the first driver to the main driver is also installed. *Bottom:* Figure 9. The brake hanger bracket is installed on the frame between the drivers. The secondary side rod connecting the main driver to the rear driver has been made and installed. The spring rigging/equalizing of the drivers is complete (both courtesy Donald S. Frost).

section would cost. But it also took longer to build and machine due to needing to fabricate the parts. With designing and fabricating the parts, it probably took two to three times longer than using existing plans and castings.

For designing, I have also done drawings to break down the full parts into components for fabrication. Each page takes an average of 1.25 hours to do. I have over 110 drawings and am maybe one third of the way through the design.

The following are some of the drawings and pictures of the engine so far. Due to the downturn in the economy, the owner has asked us to stop working on the engine.

Figure 10. The rail joint bolts are staggered so the head and nut alternate sides.

Track Basics

Track Joints

When laying track, it is a good practice to alternate which way the bolts head through the rail. Full-size railroads use this technique, because if a car or engine derails, the nut side can be sheared off and the bolts with the heads will be passed over, keeping the rail joint intact. Having a tie under the joint is helpful to support the joint, but there are times where it doesn't work out and that is fine. The joint should have an expansion gap to give space for the rails to expand as the rail heats up.

When spiking the ties, the spikes should not be straight across from each other as then the tie will be free to twist. The spikes should be staggered from the outside to inside of the rail. But they should not angle the same direction as that can cause the tie to twist and change the gauge. The rail should be spiked so the spikes that are on the inside of the rail are towards the same edge of the tie and the spikes that are on the outside edge of the rail are on the opposite side of the tie. If the tie shifts, this will help

Figure 11. The "spikes" are set so the spikes on the outside of the tie are in line with each other and the "spikes" on the inside of the rail are in line with each other. The "spikes" are set properly on the lower two ties and are set wrong on the top tie.

keep the tie from twisting. (The top tie in figure 11 is done poorly. The bottom two ties are done properly.)

The spikes (I used sheet metal screws) should be driven vertically into the tie and not driven in on an angle. The horizontal forces on the rail can help loosen up a spike that is driven in on an angle. I used 1" long #8 (in some places I used a #10) sheet metal screws as they have a nice large head to help hold the rail. Miniature spikes won't hold as well as the full-size ones do since the grain structure of the wood doesn't scale down. If a miniature spike gets loose, it can't be redriven tight. If a screw backs out and gets loose, it can be screwed in tight again.

Guard Rails for Bridges, Trestles and Curves

The guard rail should be placed inside the gauge far enough away from the main rail to allow the widest wheel to easily pass between the main rail and guard rail. If it is too tight, the wheel might not be captured but skip right over the guard rail. If a guard rail is used on a curve, it should be placed along the inside rail of the curve. This way, if a car or engine derails, it will head towards the outside of the curve and the guard rail along the inside rail will keep it from doing that.

Guard Rails on Switches

The guard rails on switches should be set so that the wheel set with the largest back to back does not pick the point of the frog, but if the wheel set is up against the guard rail, the other flange has clearance with the point of the frog. If it is not set up this way, a wider axle set could pick the frog and derail. If the guard rail is too far away from the stock rail (or is not there at all) that axle set will be allowed to wander and pick the point of the frog.

The wing rail is then set so the narrowest back to back axle set does not ride up on the wing rail.

In figure 12 the guard rail is set properly. The flange is against the guard rail and as in figure 13 the other flange is not picking the point nor hitting on the wing rail.

However, if the guard rail is missing or is too far away from the stock rail, as in figure 14, ... then the other flange is free to wander and pick the frog (figure 15). This can damage the frog and cause the equipment to derail.

Points on Switches

In full-size, the point starts below the stock rail height and catches the flange. Then the point rises so it is ¼" above the stock rail so if wheels are hollow (from being worn)

Top, left: **Figure 12. The properly set guard rail on a switch (opposite of the frog) keeps the wheel set from drifting too far.** ***Top, right:*** **Figure 13. With the guard rail keeping the axle from drifting too far, the flange does not pick the point of the frog.** ***Bottom, left:*** **Figure 14. Without a guard rail at the frog the axle set is allowed to drift.** ***Bottom, right:*** **Figure 15. Without the guard rail to keep the axle set from drifting (or with a guard rail that is set too far away from the stock rail), the axle set can drift so the flange interferes and picks the frog point. This can cause the car to derail and will damage the frog point.**

the outside edge of the wheel won't interfere with the stock rail. The point is then back at stock rail height when it connects at the heel block. In 1:4 scale it would be a good idea to have the point rise so it is 1⁄16" above the stock rail and in 1:8 scale maybe 1⁄32" above the stock rail.

The stock rail for the diverging track in full-size is also kinked right at the point (the kink depends on how sharp the curve is) and the points are straight. The frog is straight and the stock rail curves at the desired radius/curvature and the rail heading into the frog is curved slightly tighter to account for the frog being straight. The angle of the frog is a tangent to the desired radius. For each radius, the frog will have a different angle.

Ballast and Tamping

Ballast (crushed stone) is used to give a good stable base for the track, to anchor the track so it doesn't move freely out of position and to allow drainage. The ballast should have rough/sharp edges so they can interlock. Rounded stones (such as found in rivers) won't interlock very well. To allow for proper drainage, the ties need to be sitting on ballast and not on the bare ground. Once the ground is graded, a decent layer of stone should be laid down, then the track can be laid on top of the initial layer of stone. More ballast is then added on top.

When tamping, do not tamp in the center of the tie; rather tamp both sides of the outside ends of the tie and both sides of the tie just inside the rail. When tamping, use a solid bar (I use a steel bar) and pound it into the ballast near the tie. This will force the ballast under the tie to support the track. The reason the center should not be tamped is that it may be possible to pack the ballast under the center of the tie tighter than the two ends. This gives more support in the middle, causing the tie to rock and bend and raising the risk of breaking the tie. Once the ties are tamped, more ballast can be added to level up the ballast and make the track look neat.

Grades, Straights and Curves

On a full-size railroad a 1.2 percent grade (a 1.2' rise in 100', 1:83 grade) is considered a normal maximum for a main line. The steeper the grade, the less an engine can pull. If what an engine can pull on straight, level track is considered to be 100 percent, a 0.2 percent grade (2.4" rise in 100') cuts the train down to 50 percent, a 1 percent grade cuts a train down to 16.6 percent, a 2 percent grade cuts a train down to 9 percent and a 3 percent grade cuts a train down to 6 percent. (As an aside, most roads are marked as a steep hill at a 12 percent grade or more.) The biggest danger is the steeper the grade, the greater the risk of having a runaway train. Most scale equipment only has brakes on the engines and the cars don't have any. This minimizes the braking power for coming down hills, especially if the hill is on a long straight of way (curves add resistance which helps keep the speed under control when heading downhill). For most scale railroads I would recommend a 2 percent grade maximum. There are times when up to a 3 percent grade is needed, but I would use that sparingly.

On straight track, the rails should be level with each other, whereas on curves, the outside rail is angled above the inside rail to give the curve superelevation. The tighter the curve and the faster the train is to run over the curve means the more superelevation the curve needs. I have around a 50' radius curve and the superelevation is around 1⁄8"–

$\frac{3}{16}$". When transitioning from a straight to a curve, full-size railroads slowly decreased the beginning radius to the desired radius (slowly tightening up the curve) instead of going directly from a straight to a curve. This is a good practice for miniature as well; as the radius decreases, the superelevation increases. Once the track is graded, tamped and ready for service, riding an engine or car over the track will let the operator know how it feels. If it doesn't feel right, then the track may need to be adjusted. The track should feel smooth and natural and not like the car wants to tip over on the curves or is bouncing and rocking down the track.

The best practice is to use the greatest radius of curve that is possible and to use the smallest grade as possible. This will minimize the resistance that the track will have on the train.

Gauge

All of my equipment is 7.5" gauge and my track gauge is 7.530" max on the straights ($\frac{1}{32}$" over) and 7.562" max on the curves ($\frac{1}{16}$" over). In miniature and full-size the practice is to have a slightly wider gauge in the curves than the straights to allow extra room for the wheel sets, especially with long rigid wheelbases, on the curves. I do urge caution about making the gauge even wider, as the wider the gauge, the more room the wheel sets have to move laterally. With 1:8 scale equipment, the width from the outside of one flange to the outside of the other flange is about $\frac{1}{16}$" less than the nominal gauge. For 7.5" gauge equipment, this means the flange-to-flange measurement is $7\frac{7}{16}$". On a curve that is set at $7\frac{9}{16}$" there is an $\frac{1}{8}$" difference between the flange-to-flange measurement and the track gauge. If the track gauge is set at $7\frac{5}{8}$", the difference is now increased to $\frac{3}{16}$". If the gauge on straight track is set at $7\frac{5}{8}$", the equipment can move laterally/hunt by $\frac{3}{16}$", which I believe is excessive as this means that the cars can be nosing as they run down the track with the front of the car going one way and the back going the other way with up to $\frac{3}{16}$" movement. Equipment with a 7.5" gauge is meant to run on 7.5" gauge track.

Grading the Track

A transit is really useful when laying out and grading the line. Height differences in the land and grade can be measured and the grade can be calculated. Since for my track I didn't have a transit but knew the approximate height difference of my land (about 63") I knew that for my 485-foot loop I would need a 3 percent grade which would also be on a 52' radius curve. I used a string line to plot the basic curve and used a board with a level to figure out the grade. With a 48" level, I needed to place a 1" high spacer under the level at the 33" mark on the downhill side of the level. Once the level read level, I had my grade. (For a 2 percent grade, a .96" spacer would need to be under the 48" mark on the same level.)

Once the track was laid I had to grade the track and level up the rails on the straight and get the superelevation of the curves. I lay down on the ground and sighted down the rail to see where the low spots (and high spots) were and then picked up the low spots and tamped them. On the straight track, I then used a level to raise or lower the opposite rail so it was level with the first rail I tamped. On the curves I sighted down the inside rail first and then raised the outside rail to the superelevation that I wanted.

When care is taken in laying, grading and leveling track, a very good running track can be made. This will also create a much more stable and comfortable ride. Track work is a lot of work and regular maintenance will need to be done to keep the track in good condition.

Book Knowledge versus Experience

At times I have heard some heated discussions between learning from books versus learning from experienced people who had been in the hobby for quite some time. In reality there needs to be some middle ground, as book knowledge was written from experience (good and bad), and good as well as bad practices and habits can be passed on by those in the hobby. Books include a lot of background, theory and a lot of information that may not appear to have any relevance to the hobby. Sometimes those parts can be overlooked and bad practices can come from that. Book knowledge can help one distinguish between good and bad advice and practices.

For example, Dad and I were told in 1999 to remove the original swing bolster from the lead truck of the 2-8-2 and replace it with a roller that used extension springs to create the pull to guide the engine around curves. We were also told to spring the engine as an 0-8-0 and not have any equalizing going to the lead or trailing trucks. This was to get more weight on the drivers and not have as much "fighting in the springs" and allow the lead truck to swing more easily than with the original swing bolster. The people who suggested this had 20+ years in the hobby. We tried it, but the lead truck would tend to derail or would not guide the engine well and the lead drivers would derail. (As a note, this system works decently on the 2-4-0, but the 2-4-0 weighs about 290 pounds [as opposed to the 2-8-2's 900 pounds] and is a lot shorter that the 2-8-2.)

In 2014 I designed and made a new swing bolster that would do a better job of guiding the engine through curves than the extension springs and roller. This works a lot better and the engine is much more solid when running through curves. I am including pictures of the swing bolster on the 2-8-2 (figure 16) as well as the 2-4-0's lead truck with the extension springs (figure 17).

If we would have had book knowledge about equalizing and how a lead truck guides an engine through the curves, we would not have followed the poor advice to remove the prototypical system.

Figure 17 shows the truck with the roller and extension springs for the 2-4-0. In 2016 I replaced the extension springs with stronger ones and that really helped center the lead truck. When we were first doing this, we were afraid to use a stronger spring and have it not be flexible enough. That was not the case. It was too weak and didn't do a good job of having the lead truck lead the engine. This is what we were told to do for the 2-8-2 but it didn't really work well. It works with the smaller 2-4-0.

Here are some other detrimental aspects that I have seen in the hobby that don't follow book knowledge:

Firing up too quickly, especially with oil and propane burners. Firing up an engine from cold should not be any faster with oil or propane than it is with coal. Each foot of boiler diameter equals about one hour of fire-up time. Firing up too quickly can create thermal shock to the sheets as they expand and contract and can cause flues, seams and staybolts to loosen up and leak and/or crack.

Figure 16. The new swing links and swing bolster are installed on the 2-8-2's lead truck.

Cooling the boiler down too much by adding water. I have seen people drop the pressure in their boilers by up to 25 psi on a regular basis as "it will come back." This amount of cool water can thermal shock the boiler and over time cause the metal to crack.

Running with the firebox doors open. This practice allows a lot of cold air to circulate over the top of the fire and can thermal shock the flues, crown, firebox sheets and staybolts. Over time this can cause them to crack, loosen up and/or leak.

Using motor oil in a lubricator for the cylinders. Motor oil will break

Right: **Figure 17. The top view of the 2-4-0's lead truck shows the top of the roller assembly, frame and old extension springs.**

down, as it is not designed to be used in the high temperatures of steam. While it will work in a car motor, it won't lubricate steam cylinders that are exposed to 300+°F temperatures and the cylinders, valves, pistons will wear out.

Not running an engine hooked up: "with these small pistons you don't see any savings of water or fuel usage." On my engines and other engines I have run (and from what others have said) you do see a difference in water usage when running hooked up. Aside from saving water and fuel, the valve gear and rods are not beat up as much. When running in the corner, it is possible to create a pounding in the rods (which in miniature sounds like the clack of worn rod bushings but is minimized or goes away when hooked up near center) which will wear out the bushings faster. The pounding generally goes away when the engine is hooked up.

Running with water in the bottom of the glass (on level track) to "allow more space for steam." While having water in the bottom of the glass is not considered dangerous (you can still see the water) it leaves no room for error (pump stops working, pressure is too low for injector to work well) and will dip out of the glass when heading downhill (so it is impossible to know where the water level is). Steaming capacity is not just about the space above the water line, but also the volume of water ready to turn into steam. If an engine is getting a wet stack when the water is carried at a more normal level, then it might be foaming or the dry pipe is too low in the boiler or the crownsheet is too high, leaving too little distance between the water level and dry pipe.

Suppliers

Steam Locomotives

Allen Models: https://allenmodels.com/
Banning Locomotive Works: http://www.banninglocomotiveworks.com/
Godshall's Custom Machining: http://www.godshallscustommachining.com/
Keyser Locomotive Works: https://www.facebook.com/keyserlocomotiveworks/?hc_location=ufi
Little Engines: http://www.littleengines.com/
Live Steam Locomotives (Rodger Goldmann): http://modelsteamengines.tripod.com/
LocoGear: www.locogear.com
Mammoth Locomotive Works: http://www.mammothlocomotiveworks.com/
Moe's Locomotive Works: http://moeslocomotiveworks.com/
Railroad Supply: http://www.railroadsupply.com/
RMI Railworks: http://www.rmirailworks.com/

Diesel Locomotives

Cannonball Ltd.: http://www.cannonballltd.com/links.html
J&S Railcar: http://www.jsrailcar.com/
L&B Trains: https://www.lbtrains.com/
Plum Cove Studios: http://www.plumcovestudios.com/index2.html
Rail Rider Supply: http://www.railridersupply.com/products.htm
Railroad Supply: http://www.railroadsupply.com/

RMI Railworks: http://www.rmirailworks.com/
Titan Trains: http://titantrains.net/
The Train Works: http://thetrainworks.com/
Tune Trains: https://www.tunetrains.com/

Cars

Adirondack Car & Foundry: http://adirondack.x10.mx/Welcome.html or https://www.facebook.com/ADKrail/?hc_location=ufi
Cannonball Ltd.: http://www.cannonballltd.com/links.html
Godshall's Custom Machining: http://www.godshallscustommachining.com/
J&S Railcar: http://www.jsrailcar.com/
L&B Trains: https://www.lbtrains.com/
Mammoth Locomotive Works: http://www.mammothlocomotiveworks.com/
Plum Cove Studios: http://www.plumcovestudios.com/index2.html
Precision Steel Car: http://www.precisionsteelcar.com
Rail Rider Supply: http://www.railridersupply.com/products.htm
Railroad Supply: http://www.railroadsupply.com/
RMI Railworks: http://www.rmirailworks.com/
Scott's Trains: http://www.scottstrains.com/
Titan Trains: http://titantrains.net/
Tom Bee: http://www.islandpondrailroad.com/tombee/
Tune Trains: https://www.tunetrains.com/
West Bend Rail Car Shops: https://www.facebook.com/WBRCS/ or http://www.thomasjhubert.com/wbjsor10/Welcome.html

Suppliers for Other Parts

These suppliers are good places to go for gauges, burners, scale pipe fittings, boilers, injectors, lubricators, track, etc.

A&K Locomotive: http://www.livesteamcalibration.com/livesteamcalibration.htm
BryteRails: http://www.bryterails.com/
Eccentric Engineer: https://www.eccentricengineer.com/
Enterprise Plastics Inc. (switches advertised in *Live Steam and Outdoor Railroading*): http://www.enterpriseplastics.net/index2.html
Fair Weather Foundry: www.fairweatherfoundry.com
G-C Lubricants Co. (steam oil): (650) 592-1050
Godshall's Custom Machining: http://www.godshallscustommachining.com/
Green Velvet (steam oil, bearing oil): http://www.steamenginelube.com/
Harmer Steel (12 pounds/yard rail, frogs for Grand Scale railroading [generally considered 12"+ gauge]): http://harmersteel.com/
J&S Railcar: http://www.jsrailcar.com/
Kraft Industrial Supply (sight glasses, gaskets, mechanical packing): http://www.kraftindustrialsupply.com/
Little Engines: http://www.littleengines.com/
Live Steam Models: http://www.livesteammodels.co.uk/

Live Steamer Parts: http://livesteamerparts.com/
LocoParts: http://www.locoparts.biz/
Midwest Scale Rail: https://www.midwestscalerail.com/products/
Miracle Graphics & Railroad Products: http://miraclerailroadproducts.com/
Otto Gas Engine Works (piston rings): http://www.ringspacers.com/
PM Model Engines: https://www.pmmodelengines.com/
Railroad Supply: http://www.railroadsupply.com/
Ridge Locomotive Works: http://www.ridgeboiler.com/
RMI Railworks: http://www.rmirailworks.com/
Scott's Trains: http://www.scottstrains.com/
Terlyn Industries (LSB 8000 water treatment): http://www.terlyn.com/locomotive-steam-boiler-treatment/

Machinery, Tooling, Metal, Nuts and Bolts

Discover Live Steam (www.discoverlivesteam.com) has a "Buy 'n' Sell" page with a section for machinery. Sometimes machinery can be found on eBay. Contacting local machine shops could also help in finding leads on good machinery that is for sale. It is well worth the price for a good, solid piece of machinery as it will hold tolerances better than a cheap, not as solid kind.

Depending on location, there might also be a metal supplier/fabrication shop in your local area. Otherwise, check out these other suppliers.

Baileigh: http://www.baileigh.com/
Bolton Tools: http://boltontools.net/
Fastenal: https://www.fastenal.com/
Global Industrial: http://www.globalindustrial.com/?infoParam.campaignId=T97&gclid=CIG0xqPsz9MCFY-CswodghEPAA
Grizzly: https://www.grizzly.com/
Jet Tools: http://www.jettools.com/us/en/home/
Kitts Industrial Tools: http://www.kittstools.com
McMaster Carr: https://www.mcmaster.com/#
MSC: https://www.mscdirect.com/
Online Metals: http://www.onlinemetals.com/?gclid=CLrWj4qczdMCFddKDQod-o0DBw
Penn Tool Co.: http://www.penntoolco.com/
Speedy Metals: http://www.speedymetals.com/
Travers Tool Company: http://travers.com
Victor Machines (larger machinery): http://www.victormachines.com/
Wholesale Tool: http://www.wttool.com/

Books

The following are books that I have found (on eBay, Amazon and at book sales) and have in my collection.

Library of Railway Practice (Published by Railway Educational Press. It is a multi-volume set.)

Locomotive Catechism (I have the one by Grimshaw. New York: The Norman W. Henley Publishing Company.)

The Locomotive Engineman's Manual (Periscope Film LLC, 2009—reprint, W.P. James.)

Railway Training by Railway Men for Railway Men (Railway Training Institute, Ernest Cordeal.)

The Science of Railways (The World Railway Publishing Company, Kirkman. It is a multi-volume set.)

Steam Locomotive Study Course (Four volumes, Wildwood Publications, edited by William C. Fitt.)

A good place to look for some of these books or these types of books would be a local library or a college library. Many times they will have been in the collection from decades before and it takes a while to find them. While I was attending Michigan Technological University, I found several old books such as these. They had not been entered into the computerized system and the last stamp on the check-out card was the early 1950s, but unfortunately I was not able to buy them as the library wanted to keep them in case someone else wanted them. If libraries are cleaning out their inventory, some of these books may be made available to purchase.

The following books are written specifically for the hobby. Kozo Hiraoka, the author of most of these, has designed ¾" scale (3.5" gauge) steam engines and walks through how to fabricate the parts.

Building the Climax (Village Press, Kozo Hiraoka)
Building the Heisler (Village Press, Kozo Hiraoka)
Building the New Shay (Village Press, Kozo Hiraoka)
Building the Shay (Village Press, Kozo Hiraoka)
The Pennsylvania A3 Switcher (Village Press, Kozo Hiraoka)
So You Want to Build a Live Steam Locomotive (Village Press, Joseph Foster Nelson)

Other Resources

Backyard Railroading (Facebook group): https://www.facebook.com/groups/backyardrailroading/

Discover Live Steam: https://discoverlivesteam.com/

Home Shop Machinist (magazine): http://www.homeshopmachinist.net/

International Riding Scale Railroad (list at Discover Live Steam): http://discoverlivesteam.com/A-clubs.htm

Live Steam and Outdoor Railroading (magazine): http://www.livesteam.net/

Miniature railroads and clubs in the U.S. (list at Discover Live Steam): http://discoverlivesteam.com/A-clubs.htm

7.25 & 7.5 gauge Locomotives and Rolling Stocks (Facebook group): https://www.facebook.com/groups/220852414703712/

Operating a Steam Engine

Operating a steam engine is a marriage between book knowledge, practical knowledge and skills, supplemented by knowing the engine and the conditions under which the engine is operating, having situational awareness and flying by the seat of one's pants. No two engines are alike and no two trips are the same. This section will walk through the basics of operating a steam engine and why things are done in certain ways.

Basic Checklist for Operating My Steam Engines

Here is the basic checklist I use for running my engines. While this is specific to my engines, it can easily be modified and applied to other engines as well.

Before firing up:

Walk the track and remove anything fouling the track (rocks, branches, etc.).
Get cars out (or get out a few and get the rest while firing up).
Get the engine and tender out and on the steaming bay.
Connect engine and tender.
Make sure throttle is shut, Johnson bar is in center and cylinder cocks are open.

Fire up prep:

Add water treatment (LSB 8000) to the boiler and tender.
Fill the boiler with water (to about one-third to one-half a glass).
Fill tender (this can be done while firing up).
Make sure grates and arch are in place, ashpan is closed.
Add kerosene-soaked charcoal to the firebox.
Bring out extension cord and squirrel cage fan.
Set fan on stack and start it.
Light the fire.
Close valves on turret.

Lubrication:

Wipe off grease fittings before greasing so dirt is not forced into bearings.
Grease each fitting until grease starts to extrude from bearing (usually half a pump or less).

Clean oil holes/cups before oiling to prevent dirt from getting into bearing.

Oil oil holes and cups on the valve gear (link block, trunion, eccentric rod connection to link, lead and lap lever, union link, lifting link, valve crosshead), knuckle pins on side rods, oil cups on top crosshead guides and put a little oil on the bottom crosshead guides.

Fill steam oil lubricator with steam oil.

Continuing fire up:

Tend fire, add coal as needed. Break up coal as it fuses together to allow air to come through the fire. The 4-6-0 requires more raking in the back of the fire.

At or above 40 psi, open blower and remove fan. Let fan run for five minutes to cool off.

Fire up slowly to minimize stress on the boiler. A good rule of thumb is an hour of fire up time per foot diameter of the boiler from a cold start to full pressure.

Before leaving the steaming bay:

Test both injectors to make sure they both work properly.

Make sure the safety valve lifts and seats where it is supposed to.

Make sure the brakes work properly.

Blow down the sight glass by doing following procedure to make sure it works properly:

1. Ascertain the water level in the glass and open the drain valve; if there is a danger of blowing away ballast, open the valve part way. The water should empty out of the glass.
2. Close the water valve to the glass. Now steam should be blowing through the glass and out the drain valve. This shows that the steam passage way and valve is clear of any large obstructions.
3. Close the steam valve. Now nothing should be exiting the drain valve. This shows that both the steam and water valves will shut off completely. If there is either steam or water still escaping, determine from which valve it is coming and repair the valve when the boiler is cool.
4. Fully open the water valve. The glass should fill with water and water should be exiting the drain valve. This shows that the water passageway and valve is clear of any large obstructions.
5. Fully open the steam valve. The steam should blow the water out of the glass.
6. Close the drain valve. The water should return quickly to the level it was at before the glass was blown down.

Running:

Place the Johnson bar in the desired direction.

Release the brakes.

Slowly open the throttle.

Keep the cylinder cocks open a while to let the condensation blow out and the cylinders get warm.

Close the cylinder cocks (in warm weather, usually close them after 30' or so).

Hook up the Johnson bar as close to center as possible for the load. This will

(1) use less steam and water, (2) help the boiler steam better as there is less demand and (3) minimize the pounding/knocking in the rods.

Tend the fire as needed; break up the coal as it fuses together (the 4-6-0 requires more raking than the 2-4-0 or 2-8-2), add fresh coal and use the rake/poker to push the coal around to keep an even, level fire.

Don't leave the firebox door open for extended periods as that will allow "cold" outside air into the firebox and can thermal shock the sheets. Fire when working up grades only when needed to minimize thermal stress on the firebox.

Add water with the injectors as needed to keep the water between one-third to two-thirds of a glass on level track (water level will vary depending on track conditions).

It is better practice to add smaller amounts of coal and water more frequently than to add large amounts infrequently. When large amounts are added, the firebox and boiler water temperatures can drop (think of adding a large glass of cold water to a pot of boiling water on the stove. The water will stop boiling). When adding smaller amounts more frequently, a more even temperature will be held in the boiler and in the firebox (think of adding smaller amounts of cold water to the same pot of boiling water. The volume can be increased without stopping the boiling) which will help keep the boiler pressure more constant and will make her easier to run and fire

Anticipate the downgrades so to avoid a runaway.

Add more steam oil to the lubricator and don't let it run dry as then the pump will need to be reprimed (add every 1.5–2 miles for the 2-4-0 and 4-6-0; add every 1–1.25 miles for the 2-8-2 before extra reservoir was added).

If running multiple hours, re-oil and grease every two hours.

Dump ash every 1.5–2 hours for 2-4-0 and 2-8-2 and every 45–60 minutes for the 4-6-0 (smaller ashpan), depending on quality of coal to keep ash from touching the grates, which reduces the air flow, reduces fire performance and the grates overheat due to lack of air passing through them.

Shutting down:

Keep the fire in good condition to prevent holes/cold spots from letting in cold air and thermal shocking the sheets.

Bring the engine on to the steaming bay with cylinder cocks open.

Spot the engine, close the throttle, set the brake/chock/clamp on the rails, Johnson bar in center and leave the cylinder cocks open (to prevent a runaway if throttle is accidentally opened).

Open and clean out the ashpan.

Remove the grates to dump the fire (2-4-0 drops out the bottom below the mud ring, 2-8-2 and 4-6-0 are removed through the firebox door with the rake).

Finish cleaning the box and ashpan.

Wet down the hot coals.

Install grates and close ashpan door.

Clean out tubes with the brush; clean out the smokebox.

Put cars and track across driveway away.

Drain tender, use sponge to mop up the leftover water.

Blow the boiler down at 40 psi.

Put engine and tender away.

Put a light bulb in the tender tank to dry it out.

Put a light bulb in the firebox (once boiler is cool enough to not melt the cord) to keep the boiler warm when in storage and keep condensation from forming inside the boiler.

Water Level in the Boiler

Having enough water in the boiler is the most important aspect of operating a boiler. The fire temperature can be above 2,000°F, which is not quite hot enough to start melting steel but can severely weaken the sheets. The water is what prevents the sheets from overheating. Even though the fire is above 2,000°F, the sheets in the firebox are only a few degrees above the water temperature, which at 125 psi is around 350°F. In order for the water to protect the steel, it needs to be above the tube nest and crownsheet. A water glass is the only means for knowing where the water level is in the boiler.

Generally speaking, on flat level track, one-third to two-thirds of a glass of water is normal. If the water is too high, then it can be carried over into the cylinders and ruin lubrication and possibly damage the cylinders. If it is too low, the crownsheet may become bare when running down a hill. The water should freely bounce in the glass and should be higher in the glass when running up hill and lower in the glass when running downhill when compared to level track.

Figure 18. The boiler has been filled with water and is reading just under half a glass of water.

It is good practice to not cool down the boiler too much when adding water (dropping 5–7 psi max) to minimize thermal shock as the boiler temperature will drop when the cooler water is added. The larger the volume of cooler water that is added, the more the temperature will drop.

Basics of Coal Burners

The main types of fuel that are burned on scale steam engines are coal, oil and propane. Since coal is a solid fuel, it needs a surface on which to burn. This surface is called the "grates." They provide a surface for the coal to sit on and allow air to come up through gaps in the grates to allow the fire to burn. Most grates are made out of bars with spaces between them for air to pass through up to the fire.

Coal needs to be added to the fire to keep the fire thick enough that there is enough fuel to keep the fire burning and the pressure up. On full-size engines, the fireman looks at the fire and tosses each scoop of coal to where it needs to be. On many miniature engines, the engineer cannot see inside the firebox so the engineer needs to use a rake or poker to push the coal around inside the firebox to keep the fire level. A good coal fire has long, yellow flames; the hotter the fire, the brighter the flames will be.

As the fire burns and the coal reaches near the end of its life, the flames will become shorter until there are no flames on top of the coal bed. Coal should be added to the fire to keep it hot enough.

The smaller the firebox, the more likely the coal will fuse together and the rake/poker stick needs to be used to break up the coal that has fused together to allow the air to pass through the coal and allow the coal to burn better and hotter.

As the coal burns, it will produce ash. The ash will fall through the grates and collect in the ashpan. Care must be taken to not allow the ash to pile up in the ashpan and touch

Left: **Figure 19. This is a view of the 2-4-0's grates, looking up from the bottom.** *Right:* **Figure 20. The coal fire is producing long, bright yellow flames in the 6" × 8" firebox. This shows that the coal is producing near maximum heat and is in good condition.**

Left: **Figure 21. The coal fire is in its last stages of life and is not producing any flames on top of the incandescent coal bed. The firebed is still producing heat, but not enough to keep the pressure up when running. Coal needs to be added to the fire to get the necessary heat for running.** *Right:* **Figure 22. The 4-6-0's firebox is narrow and long and while it will produce plenty of heat, the back of the fire (where the draft is the weakest) needs to be raked more to keep the coal from coking over.**

the grates. If the air flow through the grates is restricted, the fire won't burn as well. The grates will also burn as reduced air flow means that the grates are not getting the cooling effect from the air. The grates can be severely damaged if the ash touches them.

The amount of coal that is needed in the firebox can vary from engine to engine and from coal to coal. The 2-4-0 likes a fire that is around 1.25"–1.75" thick, the 4-6-0 likes a fire around 2" thick and the 2-8-2 likes a fire around 0.75"–1.25" thick. While the 2-4-0 and 4-6-0 have a similar grate area (2-4-0 is 48 sq. in., the 4-6-0 is 44 sq. in.) the 4-6-0 requires more raking as the grate is only 4" wide (see figure 22) whereas the grates on the 2-4-0 are 6" wide. The fire fuses together more in the long narrow firebox of the 4-6-0 which means it needs to be raked more and really dig down to the grate to break up the coal that has fused together. If that is not done, the fire suffers, with the front burning a lot hotter than the back, and the pressure drops a lot when the throttle is opened. When the fire is raked a lot more than the other engines require, the 4-6-0 steams great even when pulling a maximum train weight up a grade.

When firing with coal, it is better to add coal in smaller amounts more frequently than in one large amount every once in a while. When coal is added to the fire, the furnace temperature will drop. The more coal that is added, the more the temperature drops. While the gases can be driven off around 800°F, they won't ignite until around 1,800°F. If the temperature drops below 1,800°F, the gases will be driven off from the coal but not ignited and will escape out the stack as dark smoke. It is best to try and keep smoke to a minimum as smoke is unburned and wasted fuel and it will soot up the tubes quickly which will reduce the steaming capabilities.

The firebox door should be kept closed except when tending (adding coal, raking) the fire. Coal needs the primary air source to enter through the bottom of the fire. Running with the firebox door open will rob the fire of the best air flow, as the air will enter by the easiest route—in through the door and not up through the fire. The air that enters through the door won't be heated up by the fire (which can burn upwards of 2,100°F) and this large volume of cold air can thermal shock the sheets inside the firebox and over time can cause them to crack and leak.

Coal needs to have a good hot fire to ignite it. To start the coal fire, some people use wood soaked in kerosene, some use paper and wood and others use charcoal soaked in kerosene. I use charcoal soaked in kerosene as it is easier to buy the charcoal than it is to chop wood into small pieces. When firing up, the fire should be kept low to allow the boiler and sheets to heat up slowly and evenly. A general rule of thumb for a boiler is that every foot equals one hour of fire up time. So an 8" boiler would take around 45 minutes to fire up by this rule. Since the volumes of the boilers can vary, so can the fire up time. An 8" boiler which holds around two gallons (RRSC 0-4-0) would have a good fire up time of 25–30 minutes where an 8" boiler which holds around five gallons (Allen 4-6-0) would have a good fire up time of 35–40 minutes.

Basics of Oil Burners

When burning a liquid, the liquid needs to be heated up sufficiently that it will vaporize and burn. Alternately, the fuel needs to be atomized and broken down into small droplets that are surrounded by air so the droplets will burn. While some engines use a pot burner, which is heated up to vaporize the fuel, most oil burners I have seen

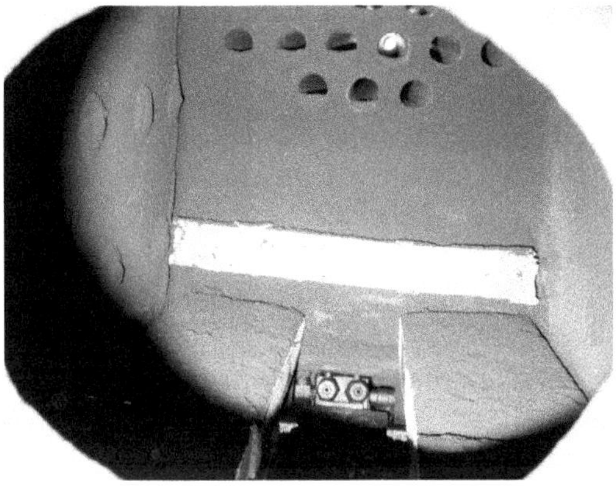

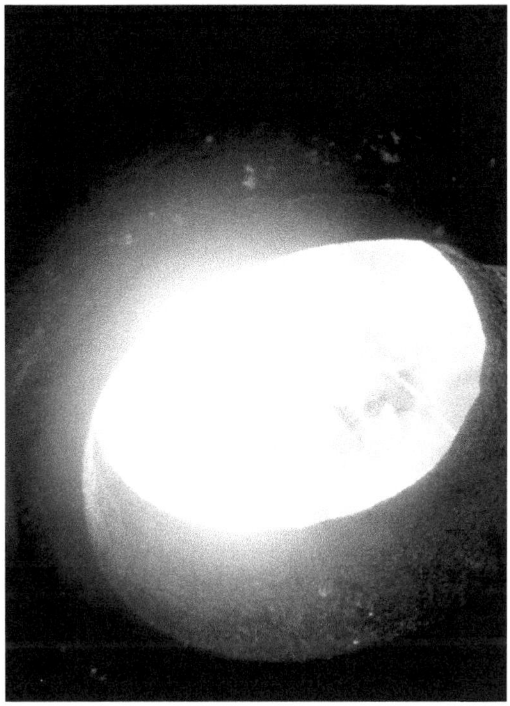

Above: **Figure 23. In the bottom center of the picture are the twin atomizers that are used on the Spirit of Traverse City, a 15" gauge oil-burning 4-4-2. *Right:* Figure 24. The bright yellow flames produced by the atomizer show that the fuel is burning well, has plenty of oxygen and is producing plenty of heat.**

use an atomizer to produce a hot, bright yellow flame. The Spirit of Traverse City has twin atomizers (figure 23).

Figure 24 shows is a "medium" fire; when the engine is working hard up a grade, the fire completely fills the firebox. The fire is lit by lighting a wick soaked in diesel fuel (what this engine burns) at the end of a long metal rod and then putting the lit wick next to the burners and turning on the atomizer valve and fuel valve. When starting an oil burner, an air compressor is needed to supply compressed air to the atomizer (and most people use it for the blower as well). The fire should be kept low to allow the boiler to warm up slowly. When changing over to steam, care is needed to not have the fire blow out.

When running and caring for an oil burner, the operator tends to be a bit busier than with a coal burner as the oil fire needs more attention and adjustment than coal. With oil, though, the fireman/engineer has a lot more control over the fire and can turn the fire up and down as needed. This also creates a risk of raising boiler pressure too fast and giving the firebox thermal shock which can loosen up flues and stress the stays, which over time can cause them to leak and/or crack the sheets. When firing up, the fire should be kept low and allow the boiler to heat up slowly. An oil burner should take pretty much the same amount of time as a coal burner of the same size to fire up. A good rule of thumb is every foot of diameter the boiler is equals one hour of fire up time (an 11–12" boiler should take around one hour to fire up).

When running, the oil supply to the burner should be turned up as the throttle is pulled out (or slightly before) because with the increased draft, more oil can be burned. If too little oil is being fed to the burner, the fire can be blown out by the draft. As the engine works harder, the fuel should be increased. As the throttle is shut and the draft is reduced, the fuel should be backed down to prevent thick black smoke. Generally speaking, a good oil burner will have a slight haze at the stack most of the time (assuming a good burner, good fuel, good draft, etc.) with a slightly darker stack when working hard.

If the draft is suddenly increased, like slipping the drivers, or a slug of water is in the steam line or the oil line gets a momentary clog, the fire can go out. If a means to relighting the fire (spark plug in the pan, firebrick on the door sheet to act as a flash wall) quickly is available, try and relight the fire. If it does not relight right away, turn off the fuel and atomizer. Once stopped, relight the fire. If the atomized fuel is allowed to build up, a spark can ignite the fire and a small explosion can result by having a lot of atomized fuel exposed to a large volume of air. Fire can come out the stack and if the firebox door is not latched, the door can blow open and the engineer can be burned. There are stories with the doors being blown off their frames on full-size engines because of this.

If the oil valve has a quadrant, it is a good idea to mark the lowest setting for the fuel (spot fire) and it may be helpful to mark a normal fuel setting and a setting for working hard up the grade. A shut off valve should be on the fuel tank and it is a good idea to have a fuel filter between the tank and the burner to prevent dirt from clogging the burner.

The atomizer valve should only be turned up as high as needed. If more steam is being fed to the burner than what is needed, it is wasting steam and the extra volume of steam in the firebox reduces the volume of air in the firebox.

When working up a grade, the blower should be able to be turned off and the draft from the cylinders should create enough draft to keep the fire burning well and pressure up. If the blower is on while working, steam is being wasted. My research indicates that for a round exhaust nozzle a good blast nozzle size for oil burners is 18 percent to 22 percent of the piston diameter.

Adjustments to the fire should be done slowly. If the engine is sitting with a spot fire, the fire should not be immediately turned up to a roaring fire as this will thermal shock the flues and firebox sheets.

Generally speaking, the firebox door of an oil burner has a damper for letting some air over on top of the fire through the firebox door and has dampers around the oil pan. The dampers are to control the air flow through the fire to create the best fire and may need to be adjusted differently depending on running conditions.

Sanding the Flues

On oil burners, when the engine is drafting the fireman will sometimes open a small door in the firebox door and pour scoops of sand into the firebox. The draft will pull the sand right off the scoop and through the tubes, knocking off soot that has built up. This is done to decrease the amount of soot build-up on the tubes as the build-up acts as insulation (it is said that 1/16" of carbon has the same insulating qualities as 3/8" of lime scale) and will cut down on the steaming capacity of the boiler. This is only done on oil burners, as oil does not have any cinders or ash that will be drawn through the tubes.

We used to sand the flues on the 2-4-0 as the ½" tubes would soot up in several hours. I made up a little scoop and made a sand box on the tender to hold the sand. We put a ½" hole in the firebox door (we tried a ¾" hole, but hole was too large and there was not enough speed to the draft in the air coming through the hole to suck the sand off the small scoop) with a sheet brass swing door. We would sand the tubes on the long grades when the engine was working decently hard. The sand was pulled off the scoop and through the tubes and the stack got darker as the carbon was knocked off. This worked quite well.

We stopped sanding the flues after installing a stainless arch in the firebox which promoted a better combustion and greatly decreased the speed at which the tubes sooted up. With the arch, we could run for five or more hours without needing to clean the tubes.

YouTube has some good videos of this process. Here is a fireman sanding the flues. You can see the sand being pulled off the scoop: http://www.youtube.com/watch?v=DzZzPTwMg98. This video is a view of the stack during sanding. You can see how dark the stack gets: http://www.youtube.com/watch?v=mUEa0r9DVwM.

Basics of Propane Burners

Propane is cleaner than both coal and oil; thus the engineer and engine will stay cleaner. Like oil, propane can give instant heat, and drastic changes to the fire should not be done in order to avoid thermal shock to the boiler and tubes. The propane tank is generally kept on a car behind the tender and the propane car is coupled to the tender with the use of safety chains (in case the coupler fails) or is connected with a solid drawbar. The connection between the tender and engine is also a solid drawbar (and the use of safety chains is a good recommendation even when using drawbars). The propane car should have an open floor (the car can be designed as a tank car, cattle car, etc.) as propane is denser than air and if there is a propane leak, the propane will drift down and it needs a way to escape.

Similar to oil, a propane fire can be chuffed out if the draft is suddenly increased, such as slipping the drivers, and a means to relight the fire should be carried (a torch works well) on the engine or tender. If the fire goes out, the propane supply to the fire should be turned off and the fire relit when stopped.

Like oil, as the throttle is opened, the fuel supply to the burners should be turned up and when the throttle is shut off/closed, the fuel should be turned down. Since propane does not produce any smoke, a pressure gauge showing the pressure to the burners is a very helpful way to see how the fire is being controlled.

Firing by Landmarks

Landmarks can be important tools in tending the fire or adding water to the boiler. Examples of "firing by landmarks" are: at XYZ crossing, cover the front half of the fire with eight scoops of coal; at the collection of old oak trees, add three scoops to the rear corners; between the collection of pine trees and ABC crossing, add water to the boiler; at D's private crossing, lightly cover the fire with coal, and so on. This style is pretty rigid and does not rely on actual, current conditions nor the understanding of how to prepare the engine for the track to come. Since each trip is never exactly the same, there is not a "one size fits all" practice.

Part of being qualified as an engineer/fireman is knowing the physical characteristics of the railroad and/or branch (grades [top of a grade, bottom of a grade, changes in grades], stations, crossings, etc.). Landmarks can be helpful. Crossing XYZ may be several minutes before the start of a climb, which makes it a good spot to start preparing the fire for the work ahead. The collection of old oak trees may be just before the start of the grade which makes it a good spot to finalize the fire before the grade. Between the pine trees and ABC crossing may be a good place to add water as a downgrade starts just past ABC crossing. D's private crossing may be a good place to lightly coal over the fire as that will keep the fire burning but not too hot for the station stop that is several minutes away.

Landmarks can also be helpful when firing as it gives time estimates to the needed demand on the engine. However, it is more important to read the fire and fire according to the present conditions (quality of the fire, quality of the coal, steam pressure, water level, etc.) than fire strictly at certain landmarks. Firing by landmarks can be used as a guideline, but should not be the final "this is the way it is done" approach. As conditions change, the fireman must adapt and fire and add water to meet the demand that will be coming.

Speed

Operating speeds of miniature trains can be a topic of hot discussion and is pretty subjective. What one person considers to be fast may not be what someone else considers to be fast. In addition, watching an engine run can make the speed look very fast, but when you are the one running it the engine seems slower. Safe operating speeds depend on a lot of things. Many scale tracks are not made to scaled-down standards for fast running either.

Many 1:8 scale cars weigh around 80–150 pounds with some weighing up to 300 pounds. When they are empty, they have a low center of gravity. Once people start sitting in the car, not only does the car get heavier but the center of gravity moves higher, increasing the risk of having a car roll over, especially when running through a curve too fast. The cars and engines that are being run can also have an effect on what the maximum safe speed is.

Superelevation on curves also affects how fast one can safely run through. The tighter the curve, the more superelevation it should have. If the outside rail is too low, the curve won't feel quite right. That, coupled with the centrifugal force acting on the riders, heightens the risk of a car rolling over.

Dips in the rails can also cause the cars to rock back and forth and create a bumpy ride. It is possible that the dips might not be felt to a great degree when running or riding, but when the track is leveled, it feels a lot better. Soft spots in the roadbed or track can sag easily and cause the rails to drop and potentially derail a car or engine. The better the track condition, the better the ride and the faster one can operate.

My roadbed is quite solid and I have around ⅛"–3⁄16" superelevation on the 52' radius curves. With an empty train I am comfortable running around 6–8 mph max, depending on the engine. With people on the train 5–6 mph max is comfortable. (If I were to let them loose, they could probably do 10+ mph up the 3 percent grade with an empty train.) I have also run on other track where 3–4 mph is too fast, as the track dips, heaves and has kinks. The 2-8-2 derailed on a stretch of straight track when the track dipped 1.5" and the engine got hung up on the lead and trailing trucks.

When I have asked people which of these to is too fast: 10 mph or 10 ft/sec, many people answer 10 ft/sec. Actually, 10 mph is faster (it equals 14.6 ft/sec). I personally haven't seen much of anything run above 6–7 mph (8.8 ft/sec–10.2 ft/sec) when people are on a 1:8 scale train. That is a lot of momentum and many cars don't have brakes. On cars that do have brakes, the brakes should be set so the wheels don't slide with an empty car (80 pounds empty car) and can weigh four times more with two adults on the car. The cars are also a lot more top heavy when people are riding in them.

Here is mph with a ft/sec equal (speeds listed in ft/sec seem faster than mph):

1 mph = 1.47 ft/sec
2 mph = 2.93 ft/sec
3 mph = 4.4 ft/sec
4 mph = 5.87 ft/sec
5 mph = 7.33 ft/sec
6 mph = 8.8 ft/sec
7 mph = 10.27 ft/sec
8 mph = 11.73 ft/sec
9 mph = 13.2 ft/sec
10 mph = 14.67 ft/sec
11 mph = 16.13 ft/sec
12 mph = 17.6 ft/sec

In March 2011 a miniature train derailed at Cleveland Park and a six-year-old boy was killed. A passenger on the train taped the train passing through a shed. It was calculated from that video that the train was traveling more than 20 mph. At a tight curve (the speed limit was around 8–10 mph for the curve, I believe), the train left the track and rolled down an embankment. The boy was killed and many people were hurt. The engineer admitted that he was going too fast. It was decided to remove the train and track from the park (http://www.rypn.org/forums/viewtopic.php?f=1&t=37148&sid=0fe4e00942630953b83e64d32bc7fbec).

Water Behavior in the Boiler

While having water in the water gauge glass and knowing that the water gauge glass is working properly are important, it is also important to understand how the water is behaving in the boiler. The water gauge glass only shows the water level at that one spot in the boiler, so it is up to the engineer to understand what this means in regards to whether or not the boiler has enough water.

Drawing 10 shows three different views of boilers: level boiler, heading upgrade and heading down grade. Each boiler has three water gauge glasses (each are the same length and located the same height from the crownsheet) drawn: rear of the boiler, middle and front. Each has the water level drawn and the water gauge glass is shaded up to the water level.

On the level boiler, each glass shows half a glass. When heading up or down a grade, the boiler will move while the water will stay level. On the upgrade boiler, the water is higher (full glass) in the rear glass, half a glass in the middle and lower (almost empty)

Level track

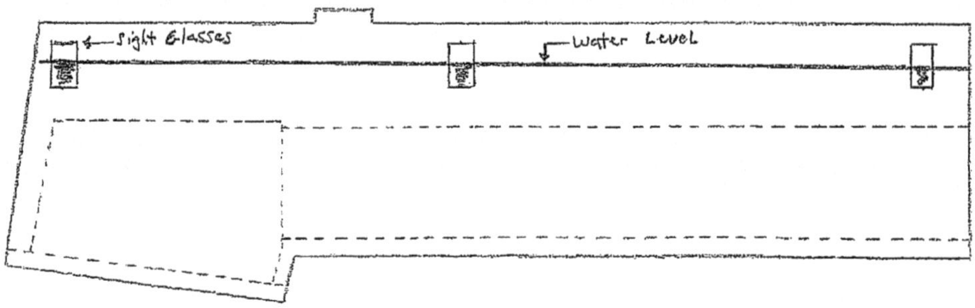

Upgrade

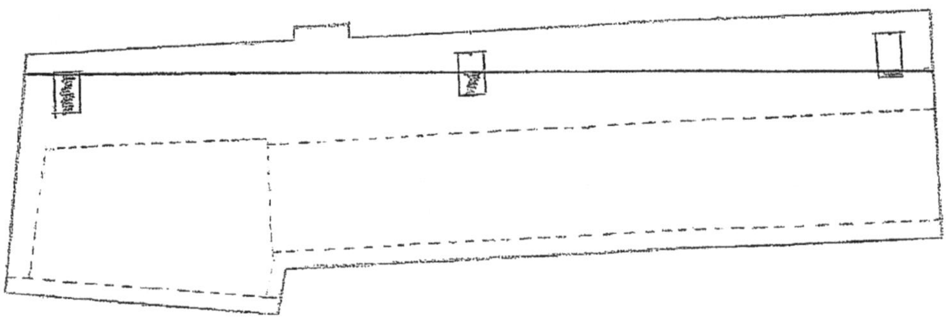

Downgrade

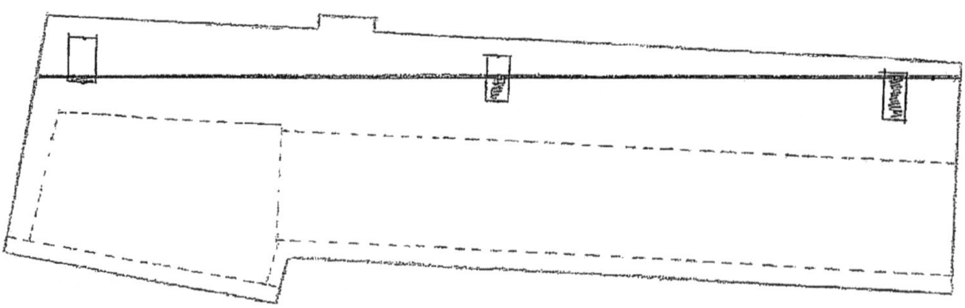

Drawing 10. The water level in the boiler will change depending upon if the boiler is sitting level, heading uphill or heading downhill. The water level in the center of the boiler doesn't vary much whereas the water level at the front and rear of the boiler varies the most.

in the front of the boiler. On the downgrade boiler, the water is lower (almost empty) in the rear glass, half a glass in the middle and higher (full glass) in the front of the boiler.

The closer the water gauge glass is to the center of the boiler, the less movement the glass will show as the grade changes.

Having a little bit of water visible in the glass while heading uphill is much more of a concern than having a little bit of water in the glass when heading downhill. If only a little bit of water is showing in the glass when heading uphill, as the engine crests the hill (or heads back down the hill) the water can drop below the bottom of the glass and then it is impossible to know where the water level is.

The longer the boiler, the more the water level will fluctuate. When cresting a 2 percent up grade, the water level will drop about ⅛" in a 25" boiler, or will drop ¼" when going from a 2 percent upgrade to 2 percent down grade. Whereas on a 50" long boiler, the water level will drop ¼" when cresting a 2 percent up grade or will drop about ½" when heading from a 2 percent up grade to 2 percent down grade.

When starting to move forward, the water will pile up at the back of the boiler and the water level will read higher on the glass mounted on the rear of the boiler until the water levels itself out. In reverse, at the start the water will pile up at the front and the water level will read lower on the glass mounted on the rear of the boiler. When applying the brakes when heading forward, the water will also pile up in the front and have a lower water level at the back of the boiler. When heading forward down a grade and slowing down, the water level in the rear of the boiler will drop even more.

Knowing whether or not the boiler has enough water depends on the following: knowing that the glass is working properly, having water in the glass, knowing how the water is behaving in the boiler (taking into consideration acceleration, deceleration, track conditions, etc.) and knowing what conditions are coming in the near future (heading uphill, downhill, slowing down, needing a greater or lesser demand for steam, etc.)

Carry Over versus Foaming

When the exhaust coming out of the stack is wet or the pop valve or whistle has water mixed in with the steam, is the boiler foaming or is the water just being carried over? They can seem similar.

Carry Over

Figure 25 shows a pot of boiling water. As the steam bubbles reach the surface, they break through the surface tension of the water. The surface of the water is not calm but is agitated by the steam bubbles rising and bursting. As the steam bubbles burst, this action can splash the water around. If in a boiler the water is carried too high (or there is minimal clearance between the dry pipe and normal water level), as the steam bubbles burst, this can splash water over into the dry pipe, throttle, turret, safety valve, whistle, etc., and produce a muffled sound with a water spray or mist as well.

If the boiler is just carrying over the water, lowering the water level will fix the problem.

Figure 25. The water in the pot is boiling and the surface of the water is not calm. The steam bubbles rise to the surface and pop. When they pop they can splash hot water.

Foaming

Left: **Figure 26. When noodles are added to the water, the starches strengthen the surface tension of the water. The steam bubbles can't pop so foam starts to form on the surface of the water.** ***Right:*** **Figure 27. The foam continues to grow on top of the water as the surface tension is too great and the steam bubbles can't pop. In a boiler, impurities and oil in the water can cause the same thing to happen.**

Figures 26 and 27 show a foaming pot of water. If the surface tension is too great (due to too many dissolved solids in the water or an oil film) the steam bubbles won't burst when they reach the surface of the water, but will rise and form foam. When a valve (whistle, throttle, safety) is opened, the pressure inside the boiler is slightly decreased causing more steam bubbles to rise to the surface, not burst and form more foam. The foam can act as a staircase for the water to climb the foam and escape through the open valve. If the safety valve is open, it won't shut at the correct pressure as the valve, is meant to work with steam and not boiling water. If the valve does not seat properly, the water may climb the staircase of foam and empty the boiler.

In extreme cases of foaming, the water in the sight glass may rise suddenly when the safety valve lifts, or when the whistle is used or throttle is opened (even when running in reverse). These actions cause the steam bubbles to agitate the water and increase the volume or raise the water level due to the steam bubbles in the water. When the valve is shut, the water level will drop (sometimes it may drop out of the glass) as the steam pressure is slightly increased and the production of the steam bubbles decreases and the water level drops and returns to a less agitated state.

If the boiler is foaming, water may be carried over when the engine is only showing ½ a glass or less of water. If a boiler is foaming, you want to limit any uncontrolled steam release (safety valve) so you don't have a situation that I listed above. You also want to carry the water level lower to reduce the chance of having the water carry over and you also want to shut off all valves periodically to let the water level drop to its non-agitated state to confirm the real level of the water in the boiler.

Oil can also increase the surface tension of the water and cause foaming. I have read that one drop of oil on water can cover a surface area from the size of a two-car garage to a couple of acres. There is not much surface area in a miniature boiler so oil can cause problems even in a very small amount. This is why oil cans or the grease gun should not be set on the tender as it would be easy to have a drop of oil or grease get into the water. It is also possible that as rubber garden hoses age they release some oil which can end up in the tender as the water passes through the hose and into the tender.

Whether the boiler is foaming or the water level is too high, water can be carried over into the cylinders and that is not a good thing. The water can wash away the oil that is lubricating the cylinders (water is not a good lubricant) and if too much water is in the cylinder, as the piston reaches dead center the water trapped between the piston and cylinder head (water does not compress) may damage or crack the piston, the cylinder head or cause damage to the running gear. The damage may not be readily apparent but may show up later.

Hooking Up an Engine

Full-size steam engines rarely run with the Johnson bar/reverse bar in the corner. When starting, the Johnson bar will be in the corner (cut-off between 75–90 percent, meaning steam is being admitted to the piston for 75–90 percent of the piston's stroke) to provide the most amount of steam (volume) to the piston and as the speed increases, the Johnson bar is hooked up closer to center so that the steam is admitted to the piston for a shorter length of time and the power comes from the expansive power of the steam and not the volume of steam. When running at speed, the cut-off may be as little as 33 percent. The boilers are not meant to supply the steam capacity needed when an engine is run in the corner and it is hard on the engine to run at speed in the corner as all the steam that enters the piston needs to exhaust from the piston and the faster the engine is running, the less time the steam has to exhaust so the engine can start to work against itself. Hooking up an engine can save steam, water and fuel which means it saves the railroads money by running the engine efficiently.

Hooking up the Johnson bar on model engines also has its benefits, but if the Johnson bar only has one or two notches near the corner, there will not be a lot of savings as there will be very little change in the cut-off. When an engine is timed well and has multiple notches to use in the Johnson bar and the engine can be run near center, the amount of coal and water used will be reduced since the cylinders are using less steam. If an engine is an okay steamer when running in the corner, hooking an engine up can help the engine steam better, as less steam is being used. The demand on the boiler is less when the engine is hooked up. Hooking up an engine can also reduce the amount of knocking or banging from the rods as less steam needs to be exhausted so the engine is not fighting against itself with trying to exhaust a large amount of steam in a short time.

However, when the track is slippery, the engine should be hooked down farther than normal to help prevent the engine from breaking loose. When the engine is hooked up, the piston is getting short bursts of high pressure steam, but when the engine is hooked down, the piston is receiving longer bursts of lower pressure steam to do the same work so the power pulses are more even.

The 2-4-0 that Dad and I built has 1.625" × 3" cylinders and was an okay steamer but became a better steamer when we added a stainless steel arch in the firebox and modified the grates to allow more air to enter the fire. Hooking her up also improved her steaming qualities as with the smaller demand for steam we did not need to add water to the boiler as often. I will generally run her hooked up near center (1 notch forward of center) unless I need more power when climbing a grade and then I will hook the engine down.

Top: **Figure 28. The 2-8-2 that Dad and I built has 2.375" × 4" cylinders and is a really good steamer. The Johnson bar (which has seven notches for forward and five for reverse) is in the corner for forward (the radius rod is down as far as it will go, the farthest away from the trunion/pivot point as it will get) which gives a cut-off of 88 percent and a valve port opening of $^{11}/_{32}$".**

Bottom: **Figure 29. The Johnson bar is hooked up two notches ahead of center (radius rod is up closer to the trunion/pivot point of the link) which gives a cut-off of 51 percent and a valve port opening of a shy ⅛". This is where she is run the most.**

Top: **Figure 30. The Johnson bar is hooked up one notch forward of center (the radius rod/ link block is just below the trunion pin and gets minimal movement from the link and eccentric crank) which gives a cut-off of 42 percent and a valve port opening of 1⁄16".**

Bottom: **Figure 31. The Johnson bar is hooked up to one notch forward of center. This notch is used on grades that are generally 1 percent or less and for the most part the 6th notch in forward (two notches forward of center) is used and we notice that the engine is less thirsty when hooked up to 6th notch versus only hooked up to 4th notch which gives a cut-off of 75 percent and a valve opening of 3⁄16".**

For comparison, when the 2-4-0 is in the corner, she has a cut-off of 75 percent and valve port opening of a strong 3⁄16". She has four notches in the Johnson bar quadrant for forward (three for reverse). When she is hooked up one notch forward of center (where I typically run her) she has a cut-off of 48 percent and a valve port opening of a shy 1⁄16".

The 4-6-0 that Dad and I built has 2" × 3.25" cylinders and I designed the Walschaert valve gear when she was built. When she is in the corner she has a cut-off of 92 percent and a valve port opening of ¼". She has six notches in the Johnson bar quadrant for forward (four for reverse) and typically I run her hooked up to one or two notches forward of center. When she is two notches forward of center she has a cut-off of 72 percent and a valve port opening of 3⁄32". When she is one notch forward of center she has a cut-off of 50 percent and a valve port opening around a strong 3⁄64". The way the castings cleaned up, the valve only has a lap of 0.080", but if the lap were greater, a shorter cut-off could be achieved.

One concern I heard about hooking up is that with less steam pressure, water may condense out of the steam and could build up as the piston approaches dead center and blow out a cylinder head.

There is nothing that supports this so it is not a concern. The valve and piston and their chambers will be pretty much at the boiler steam temperature, which at 125 psi is 352°F. If the pressure is lower than 125 in the cylinder but the temperature is the same, the steam will in effect be superheated and any condensation will be turned into steam.

If physics' law of pressure and volume is followed (P1V1 = P2V2), when an engine is hooked up to 75 percent that means that 125(.75) = P2(1). This means end pressure in the cylinder is 93.75 psi, which has a boiling temperature of 335°F, which will be cooler than the cylinder block. This means that water won't condense out of the steam as the steam is (1) at a hotter temperature than the boiling point of lower steam pressure and (2) is being exposed to a higher temperature than the boiling point of the lower steam pressure.

In July 2013 I ran the 4-6-0 while our neighbor was having a family reunion and I invited them to come over and ride the train. I ran about 12–14 trips around my 485' loop while pulling a fully loaded five-car train. I estimate that the total train weight was around 2,000 pounds. The 4-6-0 was pretty much at her limit (her calculated limit for the grade and curve is about 1,900 pounds. She can take more than this up the grade at speed, but can't start or pull more at a slow speed) going up the 3 percent grade around a 52' radius curve. Going up the hill I needed to notch the Johnson bar down to the corner or the notch just before the corner to get the power I needed to climb the hill. Here's what I take from that:

- With a heavy train and needing to run in or near the corner, I was down to 1.5" of water in the tender after 14 trips around the loop.
- With five empty cars (giving an estimated total train weight around 1,200 pounds) when I can hook her up to either notch 5 or 6 (notch 1 is in the corner) I can run 24 trips around my track before the tender is down to 1.5".
- Running with five empty cars and running in the corner, I can run around 18 trips around the track before the water is down to 1.5".

While pulling maximum tonnage does affect water usage, so does hooking up (or not hooking up) an engine. Hooking up an engine also reduces the steam demand on the boiler which means that the steaming qualities will be better as not as much steam will be used. An okay steamer can become a good steamer because hooking up an engine reduces the demand and appetite for steam.

Hooking Up, Compression and Knocking in the Rods

When an engine is hooked up, not only is the valve uncovering the port less and keeping it open for a shorter length of time, the valve is also closing the port to the exhaust passageway sooner. When the exhaust is closed, the remaining steam will be compressed between the valve and the piston as the piston moves towards dead center. On the 2-4-0, the valve closes to the exhaust when the piston is 0.340" from dead center. When she is hooked up to one notch forward of center, the valve closes the exhaust when the piston is 0.710" from dead center. When drifting down a hill when the throttle is closed and the engine is hooked up, a clacking or knocking can be heard from the rods. This is caused by the compression and momentum of the wheels being transmitted through the rods (running in the slop) to the piston to compress the remaining steam. If the throttle is opened a little, the rods will go quiet before much is heard at the stack as the steam pressure is overcoming the pressure generated by compression. To help control the negative effects of compression, the closer the engine is to drifting, the closer to the corner the Johnson bar should be.

When an engine is running at faster speeds in the corner, there is limited time for the steam to exhaust and the rods can knock or pound as they work against the back pressure from running in the corner, which will run in the slop in the rods. When running at speed, hooking up the engine reduces the back pressure and knocking (slop is not being run in) and the engine will run quieter.

Boilers

Since each state and country has its own rules, laws and codes concerning boilers, I won't be discussing designing, making, welding or repairing boilers. This chapter will concentrate on safety, water treatment, maintenance and lessons learned from a couple of accidents in the recent past.

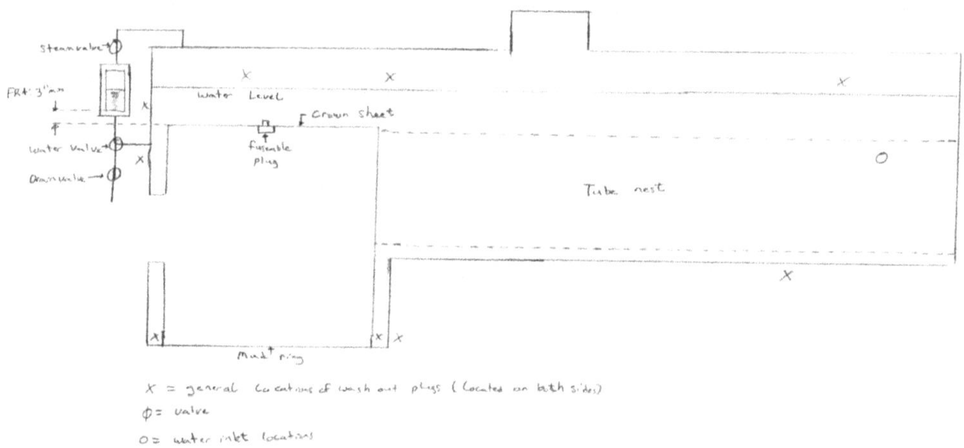

Drawing 9. The major parts of the boiler (crownsheet, tube pack, side sheet, tube sheets, etc.) are called out along with showing the water level in the boiler and it registering in the glass.

Locomotive boilers (see drawing 9) are generally constructed with the firebox at the rear of the boiler and the smokebox at the front of the boiler. The smokebox is not a part of the boiler; it is just attached to the boiler. The top of the firebox is called the crownsheet (the crownsheet on most scale models is flat whereas a full-size had the front of the crown higher than the rear), the rear sheet inside is called the door sheet, the sheets on the side are called side sheets and the sheet at the front of the firebox is called the rear tube sheet. The rear sheet of the boiler is called the backhead, the top of the boiler over the firebox is called the wrapper sheet and the sheet on the outside front of the firebox is called the throat sheet. The bottom of the firebox is called the mud right. The barrel is the cylindrical section that extends from the firebox to the smokebox. Sometimes there are several sections of the barrel with different diameters. The tubes run from the rear tube sheet (front of the firebox) through the barrel to the front tube sheet, which is the front of the boiler.

The steam dome is located on top of the boiler; sometimes it is above the firebox, sometimes above the rear tube sheet or sometimes in front of the firebox. The steam dome is above the water line to collect the driest steam available. The dry pipe runs from the steam dome through the front flue sheet. This supplies the driest steam for the cylinders.

Water Glass Safety

If a water glass appears to be clean and the water bounces freely in the glass when running, it can be very easy to assume that the water glass is working properly. Monitoring the water level in the boiler is important to the safe operation of any boiler. There needs to be enough water in the boiler to cover the crownsheet, the highest part of the firebox, to keep the metal from melting but not too much water where it could be carried over with the steam and into the cylinders. Water glasses and tri-cocks (also called gauge cocks) are normally used to monitor the water level in a boiler. The FRA requires that steam engines have two water glasses and some engines also use gauge cocks as a third way to monitor the water level. On most live steam engines, there is only one water glass in the cab and it is important to know that it is a good working condition.

With large scale and full-size engines, valves are available specifically for use with a water glass. The water/lower valve to the glass will shut off the water to the glass and to the drain valve and the steam/top valve will shut off the steam supply to the glass (and drain valve and related piping). However, with the smaller models, builders will generally use model taper pipe fittings and valves or valves that are available at the hardware store.

When making the valve assembly for the sight glass, the steam valve is attached to the top of the glass or along the piping from the top of the glass to the top of the boiler. The valve should be in an easy access area so it can be closed quickly if the glass or any of its gaskets fail. The steam supply to the glass should be connected directly to the steam space in the boiler, above the top of the glass. The steam supply for the glass should *never* be attached to the turret (which supplies steam to other devices such as the blower, injectors, atomizer, steam pump, etc.). The pressure in the turret will vary depending on how much steam is being used with the blower, injector, pump, atomizer, etc. The more steam that is being use through the turret, the less the pressure is in the turret. If the steam valve for the glass is attached to the turret it is very possible for the water pressure coming in the bottom of the glass to be greater than the steam pressure in the turret and can give a false water reading that is higher than it actually is. We found this out the hard way when we first steamed up our 0-4-0, which was later modified into a 2-4-0. We found that if the steam supply for the glass was attached to the turret and we turned the blower from off to full on, the water level in the glass would rise by at least ⅜". Once we attached the steam supply to the top of the boiler, this problem was solved.

When making the water and drain valve assembly there are a couple of choices of how to arrange the valves. One way is to bring the water supply out of the backhead or side sheet (below the bottom of the water glass and not have a trap for the water to collect in) and bring the pipe out to a tee. On the top of the tee is the valve to shut off the water supply to the glass and on the bottom of the tee is the drain valve. While this arrangement will allow the water to be shut off from the glass, I would not recommend it as (1) the steam valve will be disconnected from the drain valve when the water valve is closed so the glass cannot be blown out, (2) it would be impossible to do a proper blowing down of the glass as described later and (3) the drain valve is always connected to the boiler so if a glass needs to be replaced while under steam, there will be no way to drain any pressure that leaks past the water and steam valves while installing a new glass.

I would recommend that the water valve to the glass be located in between the boiler and the tee leading to the glass and drain valve. This way when the water valve is shut off, the water will be shut off to the glass and to the drain valve while leaving the drain valve in connection with the steam valve on top of the glass. The drain valve can bleed away any steam or water that leaks through the closed steam/water valves so the glass can be changed easily and safely while the boiler is under steam.

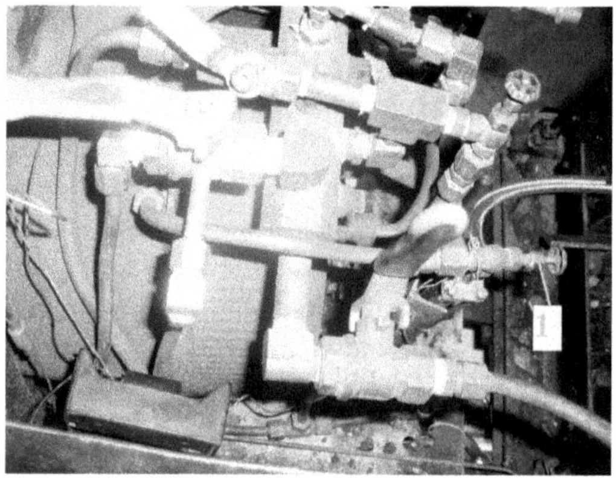

Left: **Figure 32. This is the water glass arrangement on the 2-4-0. On top of the glass is the steam valve (1). The water valve handle is seen above the drain valve, pointing towards the 10 o'clock position (2) and just above the cab floor is the drain valve (3).** ***Right, top:*** **Figure 33. This shows the steam valve (1) to the water glass and the piping to the top of the boiler.** ***Right, bottom:*** **Figure 34. This shows the water valve (2) leading out from the boiler to the tee which branches to the glass and drain valve. The piece of aluminum angle behind the glass contains three grain of wheat light bulbs to light up the glass under dark conditions.**

Blowing Down the Water Glass

1. Ascertain the water level in the glass and open the drain valve. If there is a danger of blowing away ballast, open the valve part way. The water should empty out of the glass.

2. Close the water valve to the glass. Now steam should be blowing through the glass and out the drain valve. This shows that the steam passageway and valve is clear of any large obstructions.

3. Close the steam valve. Now nothing should be exiting the drain valve. This shows that both the steam and water valves will shut off completely. If there is either steam or water still escaping, determine from which valve it is coming and repair the valve when the boiler is cool.

4. Fully open the water valve. The glass should fill with water and water should be exiting the drain valve. This shows that the water passageway and valve is clear of any large obstructions.

5. Fully open the steam valve. The steam should blow the water out of the glass.

6. Close the drain valve. The water should return quickly to the level it was at before the glass was blown down.

It does not matter whether the water valve is the first or second to be closed. The important thing is that the steam and water valves are independently tested for whether or not they are clear of any large obstructions. It is perfectly fine to close the steam valve, close the water valve, open the steam valve and then open the water valve as it is still testing what is outlined in steps 1–6; only the order is changed.

When opening the valves I have seen two ways of leaving valves open. The first is to open the valve hard against its stop and the second is to open the valve all the way and to back it off one-quarter turn. The reason to open the valve up hard against the stop is to allow the valve to be as free flowing as it can be. The reason to turn the valve back a quarter turn (which won't really reduce the flow rate through the valve unless the valve can only be opened a little amount such as one turn) is that if someone does not know if the valve is open or closed, they will be able to open the valve that quarter turn before the valve is against its stop. If a valve is open hard against its stop, if someone does not know the valve is open, they might think it is stuck closed and then use force to try and open it. Valves have been broken in that manner, allowing the valve handle and steam to be blown off under pressure.

When opening the valves on the sight glass, they should be fully open. If the steam valve is only partially open, the steam flow and pressure into the glass will be restricted and the water level will read higher than the actual level. If the water valve is only partially open, then the water flow and pressure will be restricted and the water will appear lower than the actual level.

Cleaning the Glass and Connections

Even if the water bobs up and down in the glass when running and the water returns normally during the blowing down of the glass, the water valve and pipes may still be partially blocked by scale deposits. If a boiler is using hard water, scale deposits will be

Left: **Figure 35. This is the water valve and its connections to the water glass on the Spirit of Traverse City, a 15" gauge oil-fired 4-4-2 that ran in Traverse City, Michigan, from 1982 to 2011. Hard water was being used and was being treated with LSB 8000. However, after about 250 days of service, this is the amount of scale that built up in the water passageways. If using hard water, it would be a good idea to check and clean the passageways after 30 or so days of running. If using soft water, probably every 60–90 days would be fine. Doing it more frequently won't hurt anything. Most model engines probably don't run that many days so once a year would most likely be enough.** ***Right:*** **Figure 36. Another view of the scale build-up after the Spirit of Traverse City's 250 days of service (both photographs courtesy Donald S. Frost).**

deposited faster than if softer water is being used. Water treatments can help treat the effects of scale deposits, but won't stop the deposits from happening.

While doing a small repair on the water valve on the 2-4-0's sight glass I found some pieces of scale in the valve body. While the pieces of scale were not very large, the openings in ¼" hard copper tubing and the ¼" MTP globe valve are also not very large. I did not have any problem with the water flowing freely through the pipes and glass the previous run and I doubt that those small pieces of scale that I saw would have caused any problems. However, the pipes and valves have been scaled down but the flecks of scale can never be 1:8 scale, I cleaned out the valve and the pipes as it could be relatively easy for more flecks of scale to start to build up in the narrow pipes and valves.

The inside of the glass also should be kept clean and free of scale. The glass should not have a small ID as the smaller the ID the greater the capillary action the glass will have on the water. If the ID is very small, the water will climb the glass and give a false high reading. Also, the smaller the ID, there is a higher chance of small pieces of scale clogging it. The glasses we use on our engines are ½" OD and about a 5⁄16–⅜" ID.

In June 1995 Gettysburg Railroad locomotive #1278 suffered a crownsheet failure due to low water. The locomotive had been working up grade for a number of minutes and the crew had not been using the pump or injector to fill the boiler with water while working up this grade. The crew stated that the water had been in the glass and it only fluctuated by a half inch or so. The water should have been bouncing around in the glass and if all the valves were open and passageways were clear, the glass should have been empty before the crownsheet failure. During the investigation it was discovered that the water passageways were 75–85 percent plugged by hard scale. The railroad ran with the water in the lower half of the glass most of the time and the lowest gauge cock was completely obstructed by scale, the middle one was about half obstructed and the top one was free of restrictions. The water glass also had a smaller than normal ID (thick walled glass) and there was a question whether or not the glass had a capillary effect on the water. During tests of the glass afterwards, the glass worked as well as the normal ID glass (NTSB report: https://www.ntsb.gov/safety/safety-studies/Documents/SIR9605.pdf).

It is very important that the passageways are clear to allow the water and steam to flow freely through the glass and give an accurate reading. It is also important to make sure that the steam and water valves are clean and will seal properly. If the valves leak (as determined by blowing down the glass) they should be taken apart, cleaned and lapped (or the valve body seat can be cleaned up and the Teflon on the end of the valve stem replaced) to make them seal properly. If the glass breaks, it would not be good to have the valves leaking steam and boiling hot water while the glass is being changed or not be able to shut the valves off at all if the glass breaks.

Height Above Crownsheet

The FRA requires that the bottom sight glass be at least 3" above the highest part of the crownsheet for full-size engines. In Michigan it is required that the bottom of the sight glass is at least ½" above the highest part of the crownsheet. On smaller boilers, such as our 2-4-0, when the bottom of the glass is ½" about the top of the crownsheet, the top of the glass is at the top of the steam space. LocoParts offers a shorter glass for smaller boilers. On larger boilers I prefer to have at least ¾" in between the top of the crownsheet and the bottom of the glass. With longer boilers there will be more fluctuation in the water level when heading up and down hills and for safety I prefer to have the glass up a little higher than the minimum so the rear of the crownsheet does not risk being uncovered when heading down a steep grade.

On full-size engines in the U.S. and Canada it was standard practice to have the crownsheet slope towards the rear of the engine. This means that the crownsheet in the front of the firebox is higher than it is in the rear and the height of the glass is set off of the crownsheet height in the front. This was done as when heading downhill the water will remain level and the boiler will tip down filling the front with more water and having a lower level in the rear. With the crownsheet being lower in the rear, this takes away the risk of the rear of the firebox becoming uncovered when going down grades.

In live steam models, most fireboxes have a flat crownsheet. If a boiler is 3' long and is heading down a 3 percent grade, the water will be about 0.54" lower in the glass than on level ground. If the glass is ½" above the flat crownsheet and the water was just barely in the bottom of the glass on level ground, there will be a problem heading down the 3 percent grade as the rear of the crownsheet may be uncovered. This is why on the larger and longer boilers I prefer to have at least ¾" between the top of the crown and the bottom of the glass.

Types of Glasses

The reflex glass (seen in figure 37, which has the glass is about ¾ full) is typically used on full-size engines but I believe there are some smaller types that can be used on live steam engines. The glass is a plate which is clamped onto a rear chamber. The side of the glass which is on the inside has a prism so when the water is against it, the water will appear dark whereas when there is only steam, the glass will appear lighter. Throughout use the prism will wear and the glass will darken until it is pretty much impossible

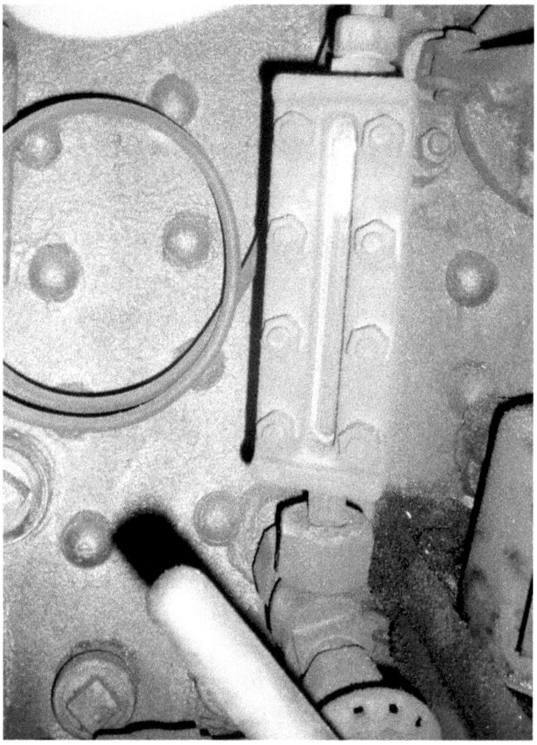

Figure 37. This reflex glass is used on a full-size steam locomotive and is showing the water level at ¾ of a glass.

to see the water level. The prism should be replaced before it is that bad. A danger with this is that if the entire glass is dark the glass could be totally full (blow down the glass to see if the color will change) or the prism is worn and the water level cannot be seen.

Most live steam models will have a tubular glass. The tube will be sealed with a bonnet nut and rubber gaskets over each end or will be placed between a top and bottom plate (with a gasket to seal the joint) and rods with nuts will be used to keep the assembly in compression (such as on the 2-4-0). Tubular glass have been known to break while in use and should have some kind of protective shield to prevent the broken glass, hot water and steam from hitting the engineer. Figure 38 shows a shield that was used on the Spirit of Traverse City. It is just a flat piece of Plexiglas sitting between the engineer and the glass.

Figure 39 shows a shield made by a friend of mine for a Cagney locomotive.

Figure 38. The Spirit of Traverse City used a tubular glass with a rectangular piece of plexiglass as a shield.

Around the glass is a semi-circle of Plexiglas and the brass cage fits around the bonnet nuts on either end of the glass and directs the steam and water towards the rear of the cage and would create a curtain of hot water over the firebox door in the event that the glass broke.

Tubular glasses will also erode. The way that they will erode and how quickly they erode depends on how the glass is held in place and how clean and/or hard the water is. It seems that the harder the water and the higher the TDS (total dissolved solids) the quicker the water end of the glass will erode. It also seems that if the glass sticks below the gaskets sealing the glass that the outside exposed to the water will erode faster than the rest of the glass. If the OD of the glass is protected by the gaskets, the ID of the water end will erode faster than the rest, but at a slower rate than if the OD is exposed to the water.

Figure 39. A Cagney uses a tubular glass with a round shield made from a Plexiglas tube fitting inside a brass housing.

Figure 40 shows how much the end of a tubular glass eroded after 30 days of service. This is the water end of the glass. The boiler was not drained once during these 30 days as it was running every day. Draining the water also helps keep the TDS levels down and appears to lessen the eroding of the glass. The glass was held in place by a rubber gasket around the glass and a bonnet nut that tightened the rubber against the valve body. The eroded part of the glass stuck below the rubber gasket and was exposed directly to the water. When a glass is held in compression and the ends of the glass are sealed by a gasket, the eroding appears to be slower, but the glass starts to thin out as the water now begins to erode at the ID of the glass, making it bigger. Due to the different conditions that live steam models face, it is impossible to give a specific time frame to know when to replace the glass as each engine will be different. Checking too frequently won't cause any harm whereas not frequently enough could cause a broken glass.

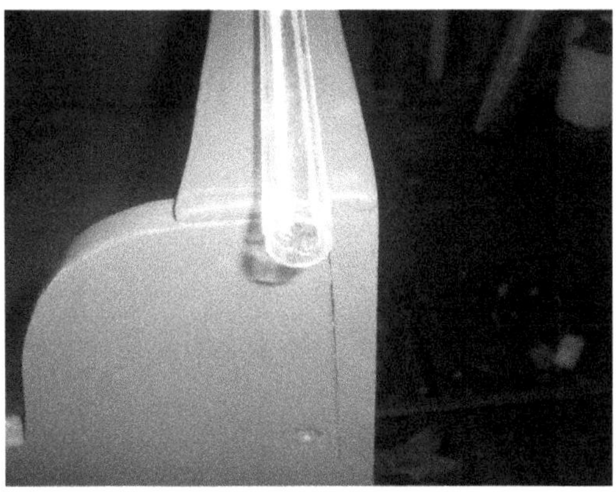

Figure 40. This tubular glass has eroded on the bottom of the water end where the glass stuck below the sealing gasket. This glass was in continuous service for 30 days.

Plumbing

When installing the plumbing for the glass, the water inlet to the boiler should be lower than the bottom of the water glass. There are times when fitting the valve on the backhead won't work due to space restraints. The water line can attach to the side sheet as long as it is below the bottom of the water glass; if the water valve is above the bottom of the glass, the line between the valve and the glass will act as a water trap and when the water drops below the water valve, the glass won't drain to show the true water level. A good rule of thumb is to keep the water line as short as possible and to have the water valve to be in an easily accessible location. The steam line should be plumbed into steam space of the boiler, above the top of the glass and the steam valve should be in an easily accessible location. The steam valve for the water glass should not be plumbed into the turret. When steam is drawn from the turret (blower, injector, water pump, air pump, atomizer, generator, etc.) the pressure in the turret will drop slightly. If the glass is connected to the turret, the steam pressure in the glass will drop slightly and the water will rise up in the glass as the water is still exposed to full boiler pressure and the glass will show a false high water level. Figures 31–34 show how we plumbed the glass on the 2-4-0. Due to the lack of room on the backhead, we had to plumb the 4-6-0's glass slightly differently (figures 41–43), using ball valves as they are easier to turn on and off than the miniature valves and due to space restraints, we had to move the valves away from the glass to areas where they would be more accessible.

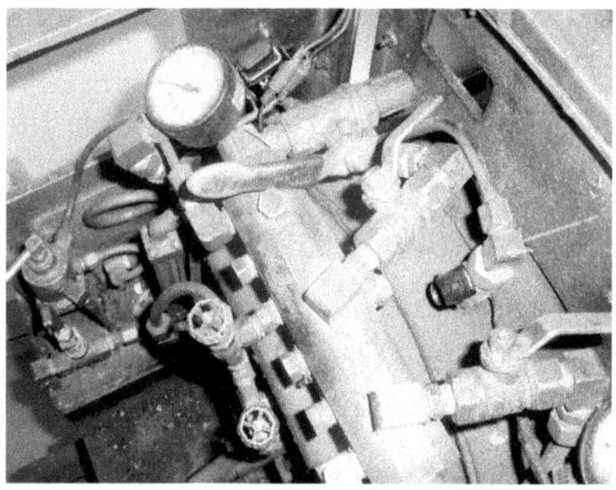

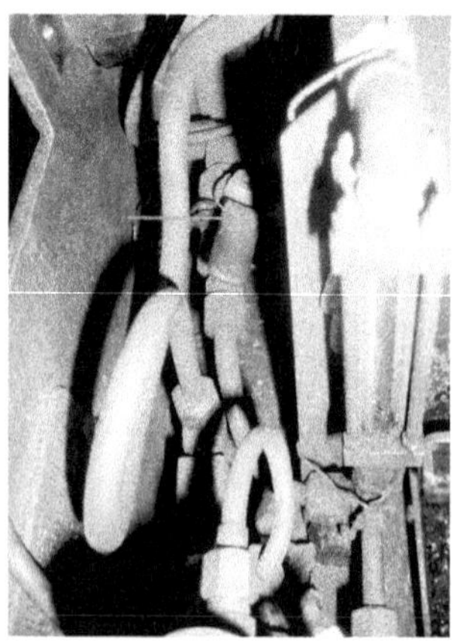

Left: **Figure 41. This is the location of the glass on our 4-6-0. The drain valve has been brought to the rear of the cab and the water line runs along the fireman's side sheet to the water valve.** ***Right, top:*** **Figure 42. The steam valve (a ball valve) comes off the top of the wrapper sheet and is easy to access.** ***Right, bottom:*** **Figure 43. The water valve (a ball valve) is plumbed into the fireman's side sheet and can be accessed through the cab window.**

Problems

Water level reads higher in the glass that it should; water glass is full and does not change. When the steam valve or passageways are partially or fully blocked/shut, the steam pressure in the glass will be less than the water pressure so the water will be forced to a higher level in the glass than what is in the boiler. If there is a leak in the steam line, the water level will read higher than the level in the boiler. The larger the leak, the higher the water in the glass will be in comparison to the water in the boiler.

Water level in glass is stationary, slowly fills up. If the water is stationary in the glass when running, this indicates that the water valve is closed and/or the water passageways are closed or plugged. The water in the glass is caused by the steam condensing and the condensation will slowly fill the glass.

Water level reads lower than it should; it's hard to get water higher in the glass. If the water in the glass is lower than the water level in the boiler and it is hard to get the water level to rise (and the injector and/or pumps are working properly) then the water valve is most likely partially closed. If the water valve is partially closed, there will be less water pressure in the glass than steam, and the steam will force the water to a lower level.

The glass breaks; how to replace a glass on a hot engine. When the glass breaks, both the steam and water valves should be closed. If the valves cannot be closed at the same time, the water valve should be closed first and then the steam valve. Steam will be spraying down from the top of the glass and this will help to keep the boiling water spraying up from the bottom of the glass from spraying as far.

When replacing a glass on a hot engine, both the steam and water valves to the glass should be closed and the drain valve should be open. Care should be taken in removing the pieces of broken glass from the open ends of the fixture to avoid fingers being cut and so that pieces of broken glass won't be left in the valves and foul them. The drain should be left open while installing the new glass to allow any steam or hot water that may leak from the steam and water valves to escape and not build up pressure. Once the glass has been tightened, close the drain valve and slowly open the steam and water valves while watching for any leaks. If everything is sealed, do a blow down of the glass to make sure the glass and valves are operating properly.

Water Treatment

It is a lot easier and cheaper to treat the water that is going into a boiler than it is to fix a boiler that has been damaged by poor water quality. Using softened water is also easier on boiler maintenance as the softer the water the fewer dissolved solids like calcium there are and the less build-up of scale there will be in the boiler. Scale build-up will reduce the transfer of heat through the firebox sheets and flues. The reduction of heat transfer means that the boiler won't steam as easily and the sheets can become overheated a lot easier as the heat will build up in the sheets instead of being transferred to the water. Scale build-up also keeps the sheets damp, which promotes pitting, rusting and corrosion of the sheets. Water also has dissolved oxygen in it and the oxygen can cause rusting, pitting and corrosion of the sheets. The feed water and boiler water should be treated for scale and dissolved oxygen.

The companies I know for water treatment chemicals for boilers are Terlyn Industries

(LSB 8000 for miniature engines), Cannonball Water Group and Malkem Chemicals. I use LSB 8000 in my boilers and use softened water as well. At the beginning of the day, I add one ounce of LSB 8000 treatment to the boiler and the same amount to the tender; if I am running a long time I add an additional one ounce to the tender. This product is meant as an easy "one size fits all" treatment even though water chemistry is a lot more involved, with needing to take account of pH, hardness, dissolved solids and measuring the amount of dissolved oxygen, oxygen scavenger and scale remover. Most hobbyists, though, won't buy the testing chemicals and equipment needed to properly test and monitor the water. Based on my experience with LSB 8000 in the 1:8 scale, it does a good job of keeping the boiler clean while not needing to buy expensive testing chemicals and testing equipment.

Figure 44. This is the tube pack of the 2-4-0. When this picture was taken she had been under steam for 18 years and had been using LSB 8000 for 15 years. The tube pack shows minimal scale build-up even though only soft water had been used in the last couple years.

LSB 8000 was used in the Spirit of Traverse City, a 15" gauge oil burning 4-4-2, which ran 100+ consecutive days a year. Two ounces were put in the boiler when the boiler was filled and two ounces were put in the tender at the beginning of the day and again halfway through the day. The boiler was drained and washed every 30 or so days. After three years, the painted lettering on the steel tubes could still be read and there was maybe 0.010" of scale build-up on the crownsheet. We did not have a source for softened water so the water was pretty hard, but LSB 8000 coupled with monthly boiler washes kept the boiler clean.

Fusible Plugs

Fusible plugs (aka "drop plugs") are a type of plugs that will fail and release pressure if they get too hot. The outside of the plug is typically made out of bronze, but the center is made out of tin, lead or a mixture of the two and is designed to melt around 445°F. The plug is located in the crownsheet (see drawing 9) and when there is water above the crownsheet, the plug is kept cool by the water. If the water drops below the crownsheet, the plug will overheat as it has lost the cooling effect of the water and the center of the plug will melt out and steam will shoot out of the plug and into the firebox. This will most likely kill a solid fire (coal, wood, charcoal, etc.) but I have heard stories on the full-size where it did not kill an oil fire and the fireman was able to push the atomizer and fire against the steam escaping from the fusible plug.

The plug is designed to fail before any harm comes to the sheets in the event of low water. The temperature of steam at 200 psi is around 380°F and most fusible plugs will melt around 445°F. This slightly higher temperature won't change the color or chemistry of the metal. It is a lot cheaper and easier to replace a fusible plug than it is to replace a boiler than has been overheated and/or failed due to low water. They should be replaced on a periodic basis as the heat and chemicals from the fire can change the tin/lead composition into something that won't melt until 1,000°F+ making it just an expensive plug and useless as a fusible plug. The two companies I know of who make fusible plugs are Flexheaters and Globe Technologies.

Boiler Anodes

Many scale boilers are made out of steel but use copper tubes instead of steel tubes. Copper tubes won't rot, rust or get oxygen pitting like steel tubes will and they are a better conductor for heat than steel. One drawback is that through a process called electrolysis the copper will erode away and give way to the steel. (This is the same way batteries work and what happens with the potato battery that we made and played with as kids.) In order to keep the copper from eroding, a sacrificial anode is installed in the boiler.

Railroad Supply sells a supply of zinc anodes. They are ¼" diameter zinc rods threaded into the end of a ⅛ NPT plug. They are threaded into the front flue sheet (RRSC boilers are built with two holes in the front tube sheet for the anodes) and can easily be changed out by access through the smokebox door. Due to my engines running 20 or so days a year, I change them out every two to three years.

Figure 45. These are the zinc anodes that Railroad Supply offers.

Boiler Washes

While we are not required to follow the FRA regulations (part 230 of the 29 CFR), it is a good reference and I prefer to follow the boiler wash schedule and the applicable aspects of the boiler washing process. The FRA requires that the boiler be drained, washout plugs removed and boiler washed every 31 service days (called a 31 service day inspection, aka 31 SDI) with a service day being a day where the boiler has pressure above atmospheric pressure *and* a fire in the firebox. With the use that my engines get, they get a boiler wash every couple of years as each engine may run 15–20 days a year.

The general location of washout plugs on locomotive boilers is as follows: in the corners of the mud ring, in the backhead above the crownsheet, in the wrapper sheet above the crownsheet and in the barrel to allow washing out of boiler mud in the bottom of the barrel (see drawing 9). With model boilers being small, the crownsheet can generally be cleaned with the washout hole in the backhead. I also added washout holes in the sides of the back to be able to wash the side sheets and mud ring and use the wand to clean the sides of the tube pack. I made the wands (one has a straight end, one has an angled end) out of ¼" copper tubing and pinched the end down to get a good stream and spray of water. I also made a hook out of ⅛" steel rod to use when cleaning out the mud ring to drag the boiler mud out the washout hole.

Figure 46. These are the wands that I use for washing the boilers. They are made out of ¼" OD copper tubing and are about 3' long. I pinched the ends down; one is straight and the other is on an angle.

When the plugs are out, they should be inspected for worn or broken threads. If the threads are worn, the plugs should be replaced. Due to the taper on the plugs, they should snug up gradually. If they are decently loose and then snug up very quickly (like in one turn or less), look at the plug and there may be a step in the taper from where the threads have been worn. If the taper plug has a step in the taper, it should be replaced. The threads in the holes should also be inspected for damaged threads and cleaned up if threads are worn, pitted or missing.

Part 230 also requires that the sight glass and gauge cock valves and passageways should be cleaned on each 31 SDI. This should also be done to the model glasses and gauge cocks to make sure that they are clean and won't plug up and not work or give a false water reading due to being constricted.

Part 230 also requires that the fusible plug (if the engine has one) be removed and inspected. If a model has a fusible plug, this is also a good thing to do and look at the threads on the plug and see if the plug has been damaged or if the plug is leaking. Due to the heat and chemicals from the fire acting on the soft lead/tin center, the chemistry of the center can change and the melting point can rise from 450°F to above 1,000°F. Because of this, it is a good idea to replace the plug on a periodic basis, such as once a year on an engine that sees a lot of use or every couple of years for engines that don't see as much use.

I made wands out of ¼" OD copper tube to connect up to the garden hose. I pinched the ends down and left one straight and angled one to be able to clean the boiler better. Figures 47–52 are pictures of the 2-4-0 receiving a boiler wash. With the amount these engines are run, I generally do a boiler wash once every several years. I remove the washout plugs (figures 47–49; also refer to drawing 9) and also removed the steam dome lid (as seen in figure 44) to be able to wash out the boiler and wash the tube pack.

Top: **Figure 47. I removed the four mud ring (two per side) plugs.**

Bottom: **Figure 48. I removed the three plugs in the backhead. One down each water leg, in a position where I can also use the wand and wash beside the boiler tubes, and one in the backhead above the crownsheet to wash the crownsheet.**

Figure 49. I removed the boiler anodes (this one was at the 9 o'clock position) to allow access for cleaning the front of the tube pack. There is an anode on each side of the tube pack.

Once all the plugs are out, I use the wand and start at the top and work the wand in through each washout hole. I rotate the want to wash off any scale or debris as I work the wand through the boiler. I use both the straight and angled wand to clean the boiler.

I start washing the boiler through the steam dome to wash off what I can reach of the top of the tube pack and top of the crownsheet.

I wash the top of the crown next through the hole in the backhead and can also wash along the top of the tube pack and wash off the top of the firebox door. I then wash through the boiler anode holes in the smokebox to wash along the side of the tube pack

Figure 50. I start washing the boiler through the steam dome to wash off what I can reach of the top of the tube pack and top of the crownsheet.

Above: **Figure 51. I wash the top of the crown next through the hole in the backhead and can also wash along the top of the tube pack and wash on top of the firebox door.** ***Below:*** **Figure 52. Once I have washed through all the other washout plug holes, I wash out the mud ring. I need to use the angled wand to wash lengthwise down the water legs (and can also use the holes in the backhead to help clean and wash out the water legs) and to wash under the tube pack and under the firebox door.**

and side of the barrel and use the straight wand to wash the rear tube sheet. From the holes in the backhead (that are in the water legs) I use the straight want to wash the front tube sheet and the angled wand to wash the debris off of the tubes and into the bottom of the barrel. I use the angled wand to also wash the sides of the firebox door and clean around the stay bolts (above and below the holes) and side sheets in the water leg.

Once I have washed

through all the other washout plug holes, I wash out the mud ring. I need to use the angled wand to wash lengthwise down the water legs (and can also use the holes in the backhead to help clean/wash out the water legs) and use the angled wand to wash under the tube pack and under the firebox door.

I continue washing until clean water is pouring out of the washout holes on the mud ring. At times I need to use a small rake (made from ⅛" diameter steel rod) to rake out pieces of scale and boiler mud. Once the boiler is clean, I inspect and retape the plugs and fire her up, run her (not all the time) and blow her down to dry out the boiler.

Water Circulation in the Boiler

The circulation of the water (not to be confused by the sloshing of the water when running down the track) in the boiler is driven by the heat from the fire. The hotter the fire, the more circulation and vice versa. The cooler water is denser and will sink to the bottom of the barrel and mud ring. As the water around firebox heats up, it rises and this draws in the water from the bottom of the barrel into the mud ring. The water will rise and move forward in the boiler where it will cool down and settle down through the tubes and into the bottom of the barrel.

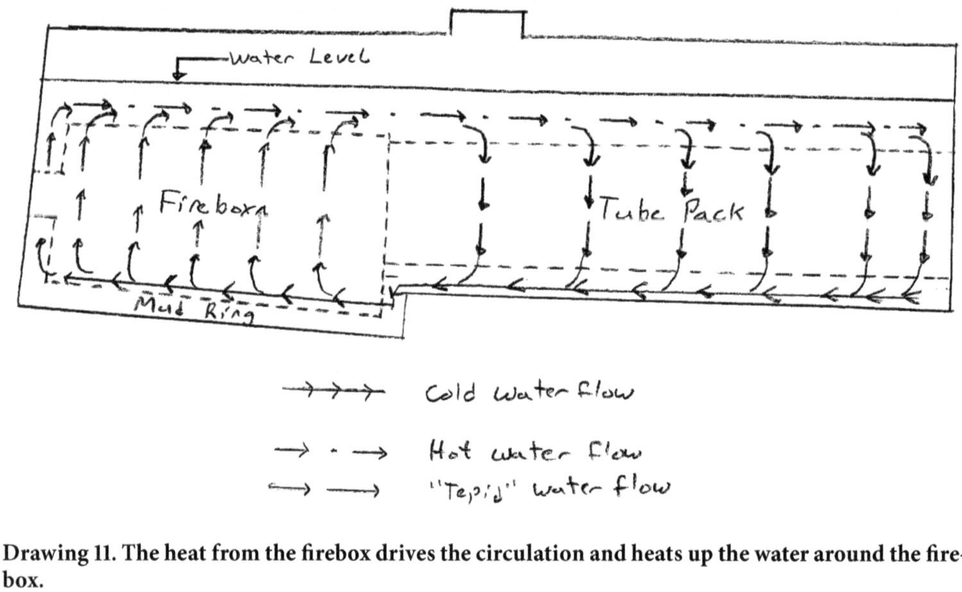

Drawing 11. The heat from the firebox drives the circulation and heats up the water around the firebox.

If a large amount of water is added at once, the large volume of cool water will collect in the bottom of the boiler and mud ring until it is heated enough to rise. This can take some time and the circulation will be greatly slowed down and there will be a lag in steam pressure until the circulation picks back up again.

If a large amount of coal is added, the firebox temperature will drop (gases will be driven off the coal around 900°F but need to be around 1800°F to ignite, so larger volumes of black smoke are quite possible) and this will cause the circulation to slow down and also create a lag in steam pressure until the circulation picks back up again.

As discussed previously, it is better to add coal and water in smaller but more frequent amounts to help keep an even firebox and water temperature. It is like having a pot of boiling water on the stove and wanting to add more water. If a full glass is added to the pot, the overall temperature drops so the pot is no longer boiling and it will take some time to return to boiling. If small amounts of water are added over a period of time, the water will remain boiling and the level can be increased without stopping the boiling.

Stored Energy in the Boiler

The boiler is made to generate steam which is then used to push the pistons back and forth; there is a lot of energy stored in the boiler. Under normal conditions the boiler contains the potential energy, but if there is a catastrophic failure, the energy can be released. During a catastrophic failure, the power that is in the boiler can peel back ⅜"–½" thick sheets of steel (weighing hundreds of pounds) like opening a metal tin can. A boiler that is at 150 psi has 21,600 pounds on every square foot of the boiler. The boiler is designed to withstand this stress. The water in the boiler is kept in a liquid state, even though it is well above 212°F, because of the pressure in the boiler: the higher the pressure, the higher the boiling temperature of the water. If the boiler pressure were to be reduced very quickly, all water in the boiler would turn to steam instantaneously. In general, one cup of water will turn into 800–1800 cups of steam. This rapid expansion is what causes a boiler explosion. The sheets in the firebox (generally ⅜"–½" thick) and in the barrel (generally around ¾" thick) can't withstand the force of this rapid expansion.

The biggest cause for a boiler explosion is that the water level was allowed to get too low and exposed the crownsheet (top of the inside of the firebox, refer to drawing 9.) Without having the water to absorb the heat, the crownsheet heats up and as it heats up it becomes weaker (as a reference, structural steel peaks in strength around 600°F. Around 800°F it is about 80–90 percent of its strength at 70°F and around 1,000°F [when it turns a dull red color], it is around 35 percent of its strength at 70°F) until it gets to the point where the force from the steam pressure bends, tears or rips the sheet and then releases the pressure inside the boiler. With the sudden pressure drop, the rest of the water boils and the result is a boiler explosion.

In general, locomotive boiler explosions are seen as a thing of the past. On June 16, 1995, a steam locomotive running at the Gettysburg Railroad suffered a crownsheet failure due to an overheated crownsheet. While it did not result in a full boiler explosion, the two firemen and the engineer (who suffered third degree burns) were burned by the steam that was released and blew back into the cab. The hot and weakened crownsheet pulled off the crown stays and formed a bulge. Steam escaped into the firebox through the empty crown stays holes. The full 63-page report can be read here:

https://www.ntsb.gov/safety/safety-studies/Documents/SIR9605.pdf.

This accident happened due to multiple reasons:

- The crew did not know who was responsible for which duties
- The crew did not know how to properly blow down the water sight glass
- The water level stayed pretty constant in the glass and moved up and down about ½"

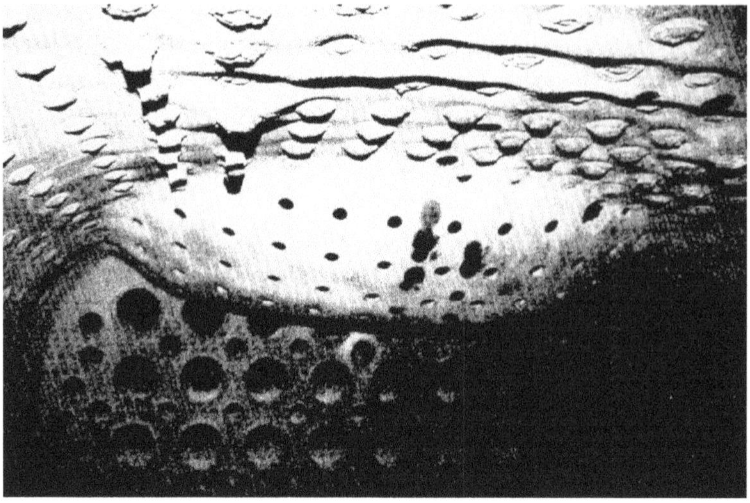

Figure 53. This is a picture of the damaged/bulged crownsheet that caused the Gettysburg Railroad accident in 1995 (page 17 of the NTSB report, found at https://www.ntsb.gov/safety/safety-studies/Documents/SIR9605.pdf).

- The crew believed that only ½" of vertical movement was normal
- The water glass and lower gauge cock were plugged with scale
- Due to a leak in the water delivery pipe, the water pump was turned off so the people on the car next to the locomotive would not be sprayed with water
- The steam valve to injector control valve was not open (injector also had the wrong nozzle so it is not sure if the injector could be used)
- The engine ran for about 10 minutes up a hill and the water level in the glass did not change
- The crew did not understand that there was a problem as the water level was not dropping even though they were using steam
- The crew was not properly trained nor had the proper knowledge/information

The lessons that can be learned from this include but are not limited to:

- Knowing how to properly blow down the water sight glass and why
- Knowing how to use and read the gauge cocks (if applicable)
- Understanding how the water moves in the boiler
- Paying attention to water and recognizing when there might be a problem
- If there might be a problem, stop running, don't ignore it
- Be trained and qualified to be an operator
- Keep the water level in the water glass at all times
- Washing the boiler and keeping the boiler clean
- Keeping the water glass valves and gauge cock valves clean
- Keep learning and refreshing oneself by reading books

Going by what is posted on Wikipedia, a boiler that holds 10,000 kg (or liters) of water (2,638 gallons) at 350 psi would have the equivalent of 1,160 kg (2,552 pounds) of TNT. The 2-8-2's boiler is quite a bit smaller at about 8.5 gallons at running level and a maximum operating pressure of 125 psi; however, if the boiler were to catastrophically fail, the energy that could be released would be about equal to the power of five pounds of TNT. Granted, not all the water will turn to steam in this event, but there is still quite a bit of power held in a small boiler. In a boiler failure, it is not just the destruction of the boiler that can be harmful. The steam can severely burn anyone near it, hot ash, cinders and coals can be blown about and the lungs can be cooked by breathing steam. Just because the boilers are small doesn't mean that they can't do some serious damage.

The operator/engineer *must* understand why water is important in the boiler and why the water must *always* cover the crownsheet and *must* keep the water level at a safe level *at all times*. While the boiler is small, it contains a lot of energy and *must* be respected, operated properly and well maintained.

Lessons from Gettysburg

The 1995 Gettysburg Railroad accident occurred due to an overheated crownsheet. Multiple things went wrong for this event to happen. While there were three crew in the cab, they did not have a clear delineation of responsibilities, the injector did not have the right nozzles and probably would not have worked with the nozzles that it had, the steam valve from the turret to the injector was turned off, the delivery line from the water pump had a leak, due to the leak the water pump was turned off for 10 minutes while running upgrade, they did not know how to blow down the glass properly, the light on the sight glass was not working, the water did not bounce freely in the glass and no one saw that as a problem, the water was not dropping in the glass even though the engine was working up grade and the water pump was shut off (no one saw that as a problem), the boiler, water glass and gauge cocks had not been properly cleaned. The three crew members were injured when the crownsheet failed with the engineer having the most amount of burns. The full report can be read here: https://www.ntsb.gov/safety/safety-studies/Documents/SIR9605.pdf.

Here is a basic list of lessons that I take from this:

1. Know how to properly blow down the sight glass
2. Clean the sight glass connections properly and on a regular basis
3. Understand the movement of the water in the boiler and how it relates to the movement of the water in the glass
4. Understand that when moving the water should be moving freely up and down in the glass
5. Understand that stationary water in the glass means that the water valve/piping is plugged and that the stationary water is the steam condensing
6. Know how to clean the boiler properly and do it on a regular basis
7. Have more than one reliable way to get water into the boiler
8. Test said ways to get water into the boiler each day to make sure they work
9. Have the proper knowledge and training (these guys didn't know what they didn't know)
10. Read books such as *The Locomotive Catechism*, *Steam Locomotive Study Course*, *Railway Training*, *The Locomotive Engineman's Manual*, part 230 of the CFR, old fireman and engineer books and/or tests to further one's knowledge above and beyond what is taught from person to person

11. Understand that if no water is being put into the boiler and the engine is being worked that the water level will drop and if it doesn't drop in the glass, there is a major problem

12. Make sure that the sight glass is reading correctly and if the engine is working without water being added the boiler, noticing that the level of the water in the glass is dropping

13. Have a working light on the water glass in miniature

Signs and Dangers of Low Water

When the water is low in a fired boiler, the sheets that are no longer covered by water can overheat. They might not fail but can leave telltale signs of damage. The sheets may be burned with the scale burning on the water side of the sheet where the sheet was exposed. Flues may start to leak; if the top row of flues is leaking, that is a potential sign that they were damaged by low water. Sheets may start to buckle and tubes may overheat and bend. Seams or joints in the firebox and/or tubes may leak and/or blow out.

It is unlikely that a miniature boiler will fail in the same way as a full-size locomotive or traction engine boiler will. Miniature boilers are overbuilt compared to their full-size counterparts. A model may have ¼"–⅜" thick firebox sheets where a full-size may have ⅜"–½" thick sheets. The thickness doesn't typically scale down. The area of the crown also does not scale linearly. If a full-size crownsheet has 80 square feet, a 1:8 scale crown will have 1⁄64th (⅛ × ⅛) the area, or about 1.25 square feet. A sheet that has a smaller area requires more force to bend it than a sheet that has a much larger area. However, if a boiler is built out of thinner materials (more scaled down steel thicknesses or thin copper) then the force required to bend the sheets is quite a bit less. Just because a boiler is unlikely to fail like the full-size doesn't mean that it should be treated with any less respect, knowledge or care.

If a boiler does fail due to low water, scalding hot water and steam can squirt from a failed seam, tube or sheet. It may be contained in the firebox, but then again it might not be. If the firebox door is open or is forced open by the escaping steam, the steam (along with any hot coals that are in the firebox) can come out the firebox door and burn the engineer. If a boiler is damaged and does not "fail" due to low water, the boiler will need to be fixed, if not fully replaced, which can give the wallet quite a hit. A compromised boiler should not be fired.

Miniature boilers should be treated with the same care, knowledge and respects as their full-size counterparts.

Valve Gear

I have designed Walschaert valve gear for a 1:8 scale engine and have rebuilt and timed Walschaert valve gear and Stephenson valve gear for 1:8 and 1:4 scale engines. This chapter will cover timing Walschaert and Stephenson valve gear, how I designed the Walschaert valve gear for the 4-6-0 and potential reasons why the valve setting might not be square and what to check. Note that the practices used for setting Walschaert valve gear can also be used for setting Baker valve gear.

Dead Centers

When setting the valves, knowing when the piston is on "dead center" is an important step. When a piston is on dead center the main rod is in line with the center of the axle so the force is directed to the center of the axle and cannot push the crank pin through its arc.

To find dead center, the piston is moved the same amount away from dead center with the crank pin going above dead center and the crank pin going below dead center. The tram bar is used to mark these locations on the driver. Dead center is between these two marks.

To find a dead center I use a tram bar, dividers, punch, hammer and a dial indicator. The tram needs to have an anchor point (punch mark) on a firm location on the engine (member attached to the frame) that does not move independently. It does not have to be any specific place and can vary from engine to engine. On the 4-6-0 I put a punch mark on the back side of the rear valve gear support.

The tram bar is nothing fancy and can be made out of round bar stock. I used ¼" diameter steel. I ground points on both ends and bent the bar so it would fit into its

Figure 54. I put the anchor mark on the back side of the valve gear support.

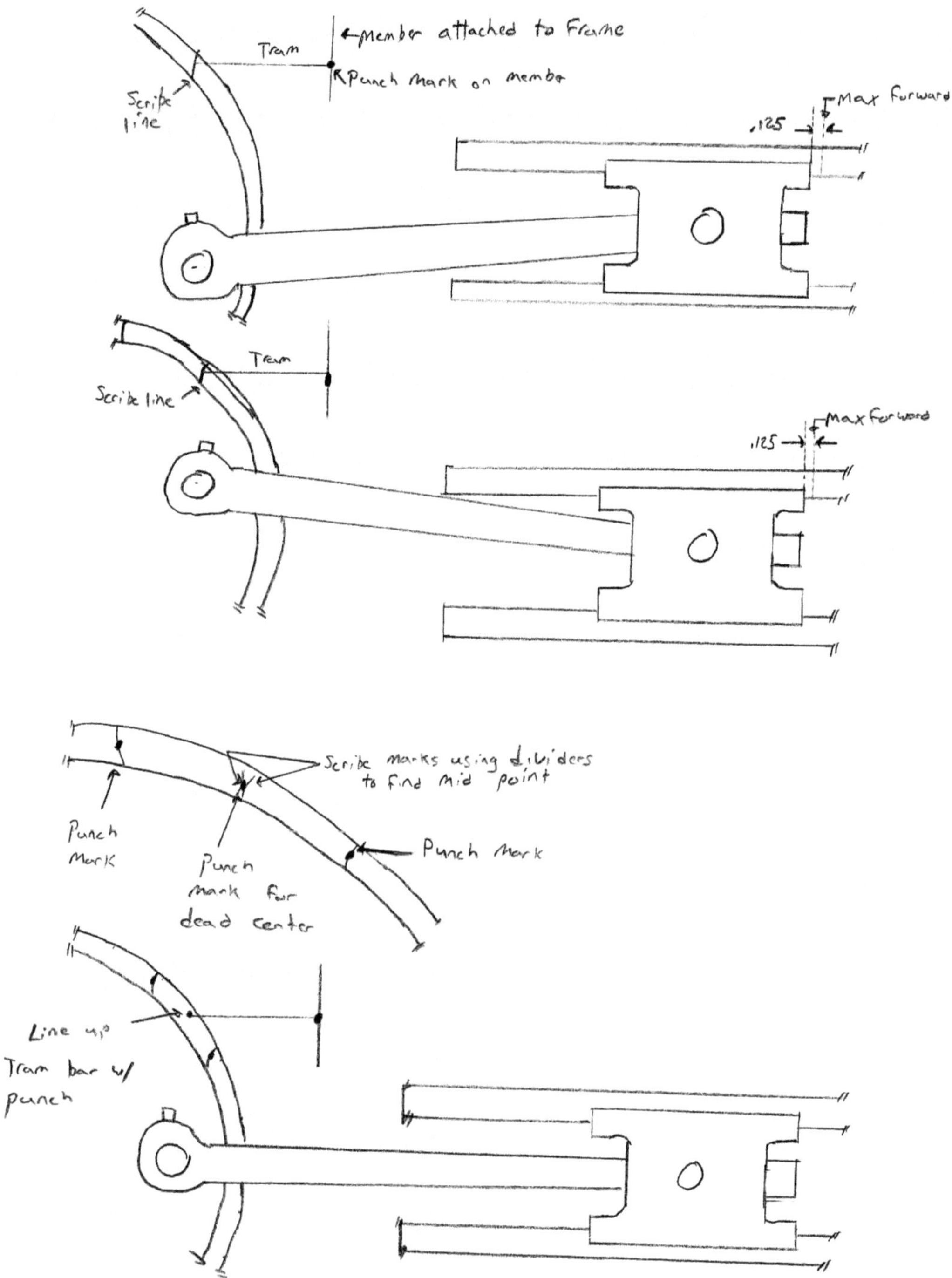

Drawing 12. The dead centers are found by scribing lines on the wheel when the piston is a set distance away from dead center. Dead center is when the tram bar distance is the halfway point between the two scribed lines.

anchor point and reach the main locomotive driving wheel tire. When the tram bar is made, it is just a piece of steel, but after using it to find dead center, it is an accurate tool. If it is dropped, bent, points mashed over, it is no longer accurate and finding dead centers needs to be repeated.

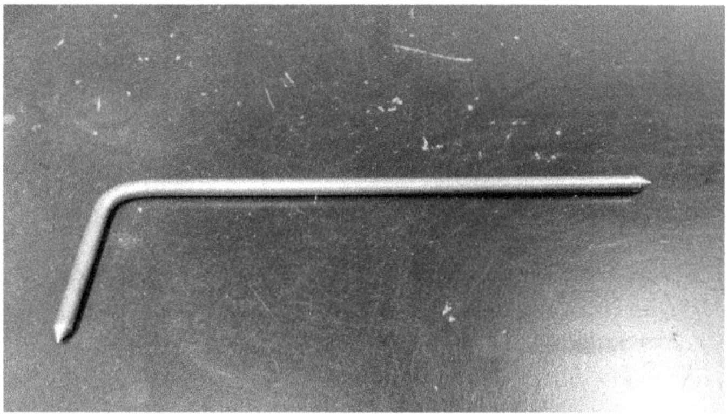

Figure 55. This is the tram bar I made to work with the 4-6-0; I have since used it on other engines.

To find front dead center, move the piston back from its extreme forward position by ⅛" (the distance does not matter. ⅛" is what I used. What matters is that the distance is accurately measured and repeated). I placed a dial indicator on the crosshead (base was mounted to the engine) to measure the movement of the crosshead. I started by having the main rod drop below front dead center. Once the piston had moved ⅛" I set one end of the tram in the anchor point and scribed a line on the tire with the other end.

I then rotated the wheels (having the piston pass through front dead center, this way the crosshead is being pulled back both times) so the main rod was above dead center and moved the piston ⅛" back from its extreme forward position. I left the dial indicator in place as I moved the engine this little bit. I then placed the tram bar in the anchor point and scribed a line on the tire with the other end.

I then put a punch mark in each of these scribed lines and used a pair of dividers to find the center between these two punch marks. I scribed two lines in the center, one from each punch mark. I then punched the intersection (center) of these lines, as when that mark is lined up with the tram bar, that will be the front dead center.

Figure 56. I am holding the tram in the anchor point and in the mark in the wheel for front dead center. The fireman's piston is now on front dead center.

The process is the same for rear dead center. I moved the piston ahead of its extreme rear position by ⅛" twice; once with the rod below dead center and once with the rod above dead center. Between these two scribed lines is dead center.

Lap and Lead

Lap and lead are terms that are used with setting the valves and are not readily understood.

Lap

Lap is how much of the valve overhangs the steam port opening when the valve is centered. Drawing 13 shows inside admission (steam enters the piston from the inside/center of the valve and from the inside edge of the port) with a spool valve and outside admission (steam enters the piston from the outside of the valve and from the outside edge of the port) with a D valve.

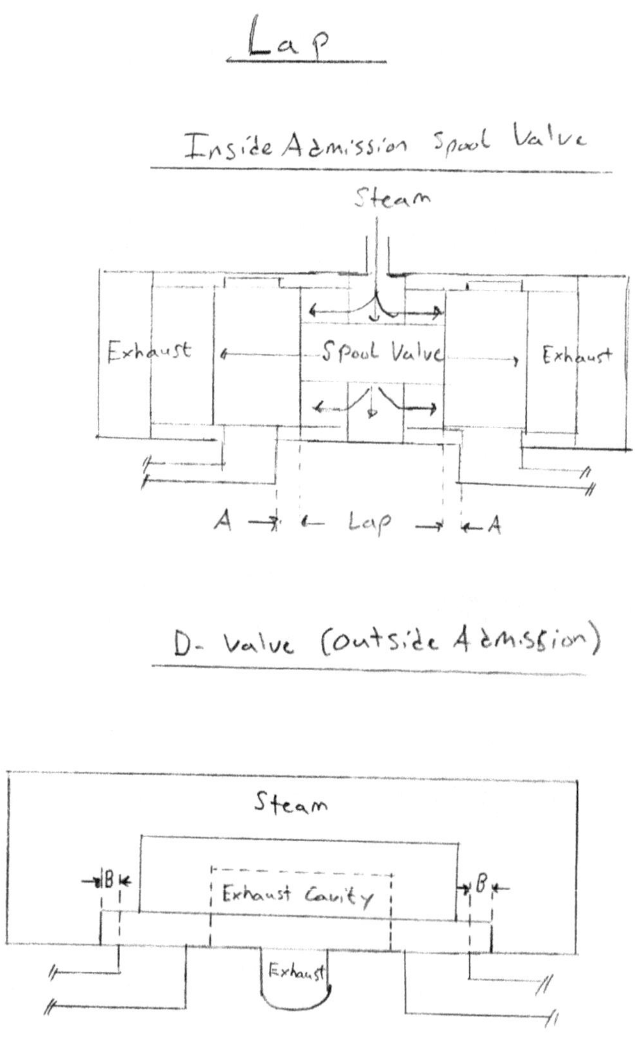

Drawing 13. The D-valve (outside admission) and inside admission spool valve are centered and the valve lap is shown.

With the spool valve, the steam is entering the cylinder from the inside edge of the port passing through the center/inside part of the valve. Dimension "A" is the lap.

With a D valve, the steam is entering the cylinder from the outside edge of the port passing by the outside edge of the valve. Dimension "B" is the lap.

Lead

Lead is how much the valve is open when the piston is on front and rear dead center. Drawing 14 shows the lead of an inside admission spool valve with a Walschaert valve gear.

The Walschaert valve gear is a constant lead valve gear so no matter where the Johnson bar is set (link block can be set anywhere in the link) the lead will still be the same.

With outside admission, the lead will let steam in from the outside ends of the valve so the outside edge of the port will be uncovered.

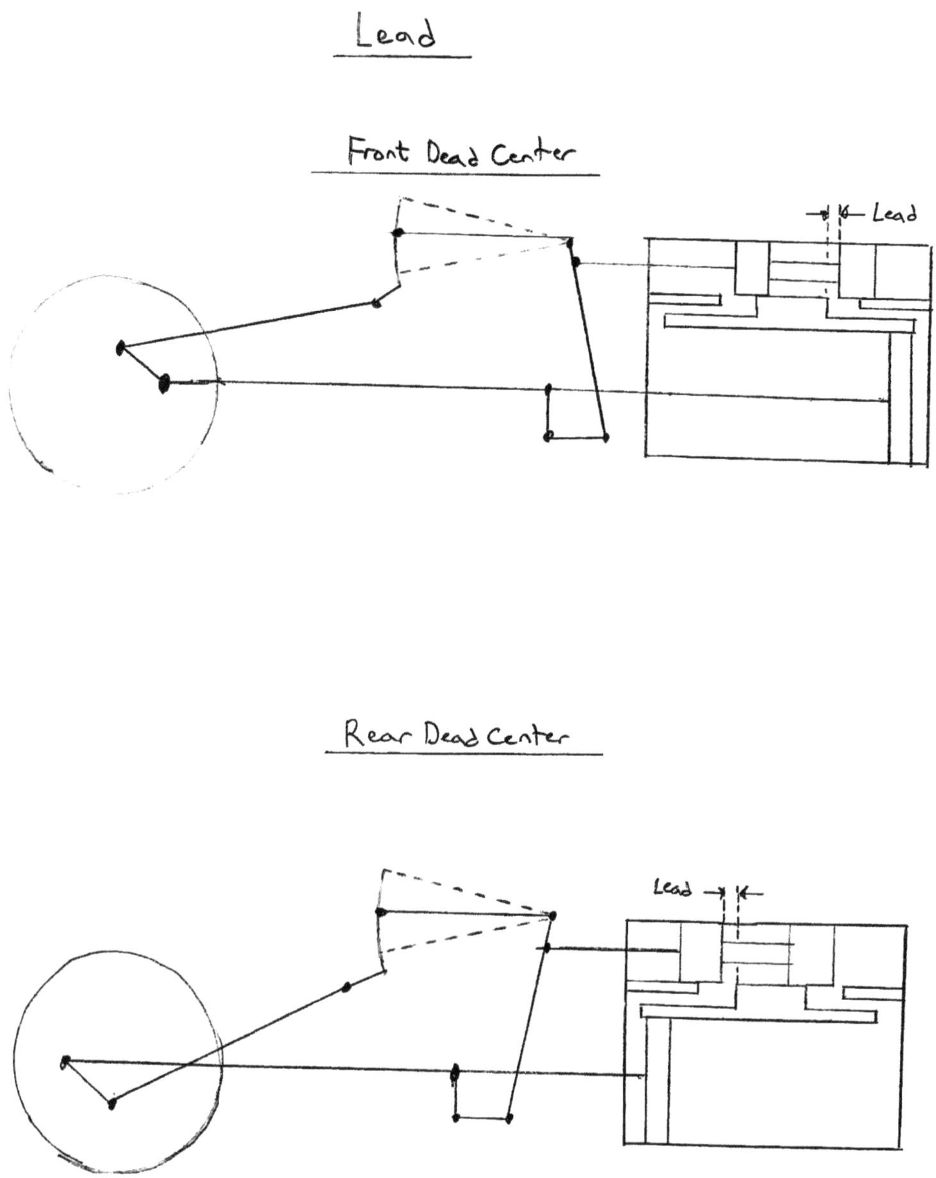

Drawing 14. The inside admission spool valve is drawn to show the lead when the piston is on front and rear dead center.

Setting Stephenson Valve Gear

Definitions

Advancing the eccentric: Moving the eccentric to increase lead at dead center. To advance the forward eccentric, rotate the eccentric forward (in the same rotation as the wheels when they move forward). Rotate the reverse eccentric backwards (in the same rotation as the wheels when they move backwards/in reverse)

Dead center: The extreme forward and back position of the piston where the main rod is in direct alignment with the axle centerline so the force applied to the piston and rod cannot rotate the drivers

Forward eccentric: The eccentric that sets the valve motion for running forward. When the engine is in the corner, only this eccentric will impart its movement to the valve. When the Johnson bar is placed anyplace in between the two corners, the movement of the valve is a combination of the two eccentrics. The forward and reverse eccentrics should have the same throw.

Lead: How much the valve has uncovered the port on dead center

Variable lead: The lead changes depending on how far the engine is hooked up. The lead with Stephenson gear will increase as the engine is hooked closer to center

Line-in-line: The edge of the valve is at the edge of the port, but has not uncovered the port

Blind: The valve is still covering the port and has to move to be line-in-line and open

Retarding the eccentric: Moving the eccentric to decrease lead at dead center. To retard the eccentric/valve, rotate the forward eccentric backwards/reverse (in the opposite direction the wheels turn in forward) and rotate the reverse eccentric forwards (in the opposite direction the wheels turn in reverse)

Reverse eccentric: The eccentric that sets the valve motion for running in reverse. When the engine is in the corner, only this eccentric will impart its movement to the valve. When the Johnson bar is placed anyplace in between the two corners, the movement of the valve is a combination of the two eccentrics. The forward and reverse eccentrics should have the same throw

Stephenson gear is typically set line-in-line or slightly blind when in the corner on the model engines. In full-size the gear is generally set with the valve blind when the gear is in the corner.

When the valve gear is applied between the frames of an engine, the link block is connected to a rocker arm that transmits the movement to the valve stem on the outside of the frame. The rocker arm reverses the movement of the link block (if the link block moves forward, the valve will move backward) so an adjustment to an eccentric rod will have an inverse adjustment to the valve. On the Shays I have seen, this is not the case as there is no rocker arm, so the movement of the eccentric is directly related to the movement of the valve.

Basic Set-up

Double-check parts against plans to make sure the dimensions are correct and no large errors are present.

Assemble the valve gear so the eccentrics are set in their approximate position and adjust the length of the eccentric rods so the rocker arm will rock evenly about a vertical line (example: from the 11 o'clock to the 1 o'clock position). The gear is not too particular for this being exact and will run well even if the rocker arm favors one side or the other of vertical.

Check to make sure that there is no binding in the valve gear; make adjustments now if there is.

Find and mark dead centers well for repeatability during the timing process.

To get the best results, replace worn pins and bushings to remove slop and reduce the lost motion in the valve gear. When setting the valves in forward, rotate the wheels for forward (and reverse for reverse) so the slop will behave the same way as when the engine is running.

Setting the Valves

With the gear in full forward (so only the forward eccentric imparts its movement) rotate the wheels in forward travel and stop at the front and rear dead centers and check the lead at each dead center. If it is uneven, make the adjustment at the valve stem so the lead at the front and rear dead center is the same (if there is 0.020" lead at front dead center, but 0.010" blind at rear dead center, adjust the valve so lead is 0.005" at front and rear dead center).

For checking to see how blind the valve is, I placed the piston on dead center and then was able to set up a test indicator on the adjustment nut of the valve stem. As I then rotated the drivers to get the valve to move, the test indicator would tell me how much the valve was moving. When the valve became line-in-line with the port, I would stop rotating the wheels and look at the test indicator to see how much the valve moved.

Once the lead is even on front and rear dead center, the eccentric will need to be moved or rotated to get the desired lead. (Using the above example with 0.005" lead on the dead centers, the eccentric will need to be moved to decrease the lead to line-in-line to 0.005" blind. In order to retard the eccentric in this manner, the eccentric will need to be rotated in the opposite direction of the engine travel, in this case rotated backwards a little bit.)

Once the lead is even and set to the desired amount, lock the eccentric in place and move on to the backup/reverse eccentric. Since the valve is adjusted properly for forward, the valve stem should not be adjusted when setting the valve for reverse.

Set the Johnson bar in full reverse and find the lead at each dead center with rotating the wheels in reverse. The lead most likely will not correlate with forward at this point. Once the lead is determined, the lead will need to be evened up between front and rear dead center. This is *not* done with the valve stem, as the valve is set for forwards. This adjustment is done by adjusting the length of the backup/reverse eccentric rod. If the rod is threaded into the eccentric strap, the strap can be threaded on or off a little at a time to adjust the position of the valve. If the eccentric rod is bolted to the strap, the holes might need to be milled out so they are oval. When the bolts and nuts are tightened, they should be tightened enough so the rod won't move in the strap.

For example, if in reverse the valve is 0.025" blind at front dead center and 0.005" blind at rear dead center, the valve will need to move back 0.010" to have it be 0.015" blind at both dead centers. Because the valve moves in the opposite direction of the link due to the rocker arm, the backup/reverse eccentric rod will need to be lengthened by 0.010".

Once the reverse eccentric rod is adjusted so the lead is even, secure the rod to the eccentric strap. The next step is to rotate or move the backup/reverse eccentric so the lead in reverse is the same as in forward.

In this example, forward was set to 0.005" blind and reverse is 0.010" blind. The backup/reverse eccentric will need to be advanced (moving the backup/reverse eccentric in the reverse direction/backup wheel rotation) to increase the lead to 0.005" blind, then the eccentric should be secured so it won't move.

At this point, lead should be equal in forward and reverse. The next step is to hook up the Johnson bar and check the lead when the engine is hooked up. If the valve gear is properly designed and machined, the lead will be even on front and rear dead center and will be increased the closer to center the Johnson bar is.

If for some reason the valve gear does not produce this result (which happens), set the Johnson bar in the "company notch" or in the normal hooked-up notch for running and set the lead equal in that notch by adjusting the valve stem. This will allow good running in the normal hooked-up notch. Running in the corner is more forgiving and flexible with sounding square and not having the valve movement being spot on.

Once the lead is set, the next step is to check the cut-off in the corners and hooking the engine up. In each notch, watch the valve until it closes the steam port. Then measure how far the piston moved. To figure out the cut-off, divide the piston movement by total stroke, then multiply by 100. The cut-off should be reasonably close (within a couple percent) for the front and rear side of the piston.

Example: In the corner, an engine that has a 4" stroke has the valve cut of steam supply at 3.610" on the front and 3.640" on the back. The cut-off is 3.61/4 × 100 = 90.3 percent for the front side of the piston and 3.64/4 × 100 = 91 percent on the rear side.

If this engine, when hooked up to one notch forward of center, has the valve cut-off steam supply at 1.990" on the front and 1.960" on the back, then the cut-off is 1.99/4 × 100 = 49.8 percent on the front and 1.96/4 × 100 = 48.5 percent on the rear side of the piston.

If the cut-off greatly varies from the front side to the rear side of the piston, the culprit may be the attachment to the link that connects the link to the lifting arm/link. If centerline of the lifting link attachment is in line or ahead of the centerline of the link block, the cut-off can vary greatly. The centerline of the lifting link attachment should be behind the centerline of the link block.

Once one side of the engine is set, the process begins on the other side. The lead should be the same forward and reverse and should be the same from one side of the engine to the other.

Setting the valve can be time consuming and frustrating, but it can also be very rewarding. If care is taken in measuring the valve movements and adjustments, a good running engine will be the end result (assuming the valve and the passageways are properly machined).

For a Shay/Clishay Where the Rocker Arm Is Not Used

While I have not seen any documentation about setting valves on a Shay, these are the differences I can see in setting the valve. Because a rocker arm is not employed, the movement of the eccentric is directly imparted to the valve. For the basic set-up, I would suggest setting the length of the forward and eccentric rods to the same length.

Like the above directions, first set the valve for forward and have the Johnson bar set in full forward and even up the lead by adjusting the valve stem. For adjusting how much lead, because the rocker arm is not employed, I would wonder if to advance the lead the forward eccentric would have to be rotated to reverse and to retard the valve the eccentric would have to be rotated forward (opposite of when the rocker arm is employed). Once the lead is even and set to the desired amount, secure the valve stem and forward eccentric.

For setting the valve in reverse, place the Johnson bar in full reverse and check lead at the dead centers. To even up the lead, adjust the backup/reverse eccentric rod. If the valve needs to move up to even up the lead, the eccentric rod would need to be lengthened, etc. To set the lead to the same amount as forward, the backup/reverse eccentric will need to be rotated.

Note: On steam locomotives the eccentric rods are set as "open rods" so the lead will increase as the engine is hooked up. (Steam ships have the rods set as "closed" so the lead greatly reduces as the valve gear is hooked up.) When Stephenson gear is in the corner (full forward or full reverse), the valve motion is controlled by one eccentric. As the valve gear is hooked up, the valve gear gets some of its movement from the other eccentric and the lead increases. In center, the motion of the valve is an even combination of the go-ahead and back-up eccentrics and the lead is the greatest in center. Because of the varying combinations of motion from both eccentrics it is possible to have an engine running forwards while the valve gear is hooked slightly behind center; the back-up eccentric would only be giving slightly more motion to the valve than the go-ahead eccentric.

Setting Walschaert Valve Gear

Here is the basic procedure I use for setting the valves on the engines. If assembling the gear for the first time, it is a good idea to check the dimensions of the parts against the plans to check for any large errors. Chances are slight adjustments will need to be made in the process, but that is typical. To get the best results, replace worn pins and bushings to remove slop and reduce the lost motion in the valve gear. When setting the valves in forward, rotate the wheels for forward (and reverse for reverse) so the slop will behave the same way as when the engine is running.

Basic Set-up

Assemble the union link, lead and lap lever (also called the "combination lever"), valve crosshead, valve stem, radius rod, link, link block, trunions, lifting link and tumbling shaft. Set the Johnson bar in center and adjust the length of the reach rod (or lifting links) so each link block is in the center of the link/trunion pin. When it is in center, the link should be able to move without moving the valve.

Setting the Valves

Before setting the valves, the dead centers should be found. If the lead is known, the lead can be set at each dead center with the Johnson bar in center. The Walschaert valve gear has a constant lead so the lead will be the same no matter where the link block is in the link. With the engine in center, the only thing controlling the movement of the valve is the piston through the lead and lap lever. Adjusting the valve stem adjusts the placement of the valve and the valve should be open the same amount on front dead center as it is on rear dead center. When rotating the wheels, it does not matter if the

Figure 57. The parts of the Walschaert valve gear: (1) lead and lap lever; (2) union link; (3) radius rod; (4) link and link block; (5) trunion and trunion pivot; (6) eccentric rod; (7) eccentric crank.

wheels are rotated forwards or backwards, but do it the same way for front and rear dead center so the slop behaves the same way.

Once the lead is set it is a good idea (with the engine in center) to turn the wheels so the piston is halfway through its stroke. The valve should now be centered (this can be seen on the spool valves on the 2-4-0 and 2-8-2, but not on the slide valves on the 4-6-0).

Setting the Eccentric Crank

The eccentric crank is a crank that is connected to the crank pin and has an offset pin to create another motion for the valve gear. The motion from the eccentric crank combines with the motion from the piston (through the lead and lap lever) to move the valve to let steam into and out of the piston chamber. If the eccentric crank is not set properly, the valve events will be off and the engine will be lame or out of square.

The eccentric crank should be set so that the eccentric crank pin is about 90 degrees off the crank pin. If the valve gear is set up for direct motion (forward in the bottom of the link, valve movement is the direct motion from the eccentric rod) for inside admission, the eccentric crank is to follow the crank pin (crank pin reaches dead center before eccentric crank pin does when running in forward).

If the left main driver has the crank pin at the 6 o'clock position, the eccentric crank pin will be around the 9 o'clock position. The 2-4-0 and 2-8-2 are direct motion for inside admission.

If the valve gear is set up for direct motion for outside admission, the eccentric crank pin will be set to lead the crank pin. If the left main driver has the crank pin at the 6 o'clock position, the eccentric crank pin will be around the 3 o'clock position. The 4-6-0 is direct motion for outside admission.

While the full-size uses a method which uses tram bars to set the position of the

Top: **Figure 58. With inside admission and a direct motion Walschaert valve gear, the eccentric crank will be set so the eccentric crank pin follows the crank pin (meaning that when running forwards, the crank pin will reach the dead center before the eccentric crank pin). On the left-hand side (fireman's side) of a locomotive, with the crank pin at the 6 o'clock position, the eccentric crank pin will be around the 9 o'clock position.** ***Bottom:*** **Figure 59. With outside admission and a direct motion Walschaert valve gear, the eccentric crank will be set so the eccentric crank pin leads the crank pin (meaning that when running forwards, the eccentric crank pin will reach the dead center before the crank pin). On the left-hand side (fireman's side) of a locomotive, with the crank pin at the 6 o'clock position, the eccentric crank pin will be around the 3 o'clock position.**

eccentric crank, I have used several easier methods. With the crank pins at the 6 o'clock position, I have eyeballed the position of the eccentric crank so the center of the eccentric crank pin, the center of the axle and the center of the crank pin form a right triangle. I have also used a combination square held up to the wheel to aid in getting the correct position. I have also drilled the eccentric crank pin hollow to insert a small welding rod through the pin to find the correct location and throw from the centerline of the axle.

Setting the Eccentric Rod

When the pistons are on dead center, the Johnson bar should be able to be moved from full forward to full reverse without the valve moving at all if the gear is set up correctly.

Above, left: **Figure 60. When the radius rod is machined to match the radius of the link, and the eccentric and eccentric rod are properly set, then when the engine is on dead center (front or rear) and the valve gear is moved through its full motion, the valve won't move. The valve gear is set in center at rear dead center.** ***Above, right:*** **Figure 61. With the engine at rear dead center and the valve gear moved to full forward, the valve did not move.** ***Left:*** **Figure 62. With the engine at rear dead center and the valve gear moved to full reverse, the valve did not move. This shows that the eccentric and eccentric rod are properly set.**

Next step is to repeat with the piston on front dead center. If the valve doesn't move again, the set-up is good. If the valve moves, then the eccentric rod and/or the eccentric crank will need to be adjusted. This is covered under the "Potential Problems" section of setting the valve gear.

I then place the Johnson bar in the corner (I typically start with forward) and rotate the wheels (forward) and watch the valve. If everything is set up correctly, the valve should open evenly on the front and rear ports, in full forward and full reverse and any place in between. (Note: with Baker valve gear, the valve doesn't open evenly between the front and rear ports and doesn't have the same cut-off between the front and rear ports. That is the nature of the valve gear.) When the Johnson bar is in forward, I rotate the wheels forward to keep the slop behaving as it will when it is running. When the Johnson bar is in reverse, I rotate the wheels in reverse.

If some dimensions are off in the valve gear, the valve movements may be off. If that is the case, set the valve for even movement where the engine is run the most (generally it is called the "company notch" where the Johnson bar is set so it is in the forward part of the quadrant, but hooked up near center). Setting the valve so it has even movement near center means that the engine will run well hooked up and running in the corner will be off some, but due to the large movement of the valve in the corner, it is very forgiving in the corner and it is quite possible that the error won't be heard.

If the valve movement looks good in forward and reverse, the valves are set.

Determining Percent Cut-off

Once the lead is set, the next step is to check the cut-off in the corners and hooking the engine up. In each notch, watch the valve until it closes the steam port. Then measure how far the piston moved. To figure out the cut-off, divide the piston movement by total stroke, then multiply by 100. The cut-off should be reasonably close (within a couple percent) for the front and rear side of the piston.

Example: In the corner, an engine that has a 4" stroke has the valve cut-off steam supply at 3.610" on the front and 3.640" on the back. The cut-off is $3.61/4 \times 100 = 90.3$ percent on for the front side of the piston and $3.64/4 \times 100 = 91$ percent on the rear side.

If this engine, when hooked up to one notch forward of center, has the valve cut-off steam supply at 1.990" on the front and 1.960" on the back, then the cut-off is $1.99/4 \times 100 = 49.8$ percent on the front and $1.96/4 \times 100 = 48.5$ percent on the rear side of the piston.

Potential Problems

- If the engine is on dead center and the valve moves forward when the Johnson bar moves forward and moves back when the Johnson bar moves to reverse, and this happens at front and rear dead center, then the eccentric rod is too long and needs to be shortened.
- If the engine is on dead center and the valve moves back when the Johnson bar moves forward and moves forward when the Johnson bar moves to reverse, and this happens at front and rear dead center, then the eccentric rod is too short and needs to be lengthened.
- If the engine is on dead center and the valve is too far forward in forward and too far back in reverse, but on the opposite dead center the valve is too far back in forward and too far forward in reverse, then the eccentric is not set correctly. Once the valve is consistently too far one direction in forward and too far the other direction in reverse, then the eccentric is set, but the eccentric rod is not.
- If the valve is set correctly in center, but moves forward when the Johnson bar is put into forwards *and* moves forward when the Johnson bar is put into reverse, then the radius of the link is too small for the length of the radius rod.
- If the valve is set correctly in center, but moves back when the Johnson bar is

put into forwards *and* moves back when the Johnson bar is put into reverse, then the radius of the link is too large for the length of the radius rod.

- If the cut-off is quite a bit uneven from the front side of the piston to the rear side, the problem may be in the position of the lead and lap lever and the length of the union link. When the valve gear is in center and the piston is centered in its stroke, the valve should also be centered in its stroke *and* the lead and lap lever should be vertical (set by the length of the union link) to allow even swinging from vertical and even valve movement forward and behind center.
- If one side has more or less lead than the other, the problem may be that the positions of the holes on the radius rod (generally speaking, where the radius rod attaches and valve crosshead attaches) is not the same spacing from side to side which allows the different lead between the sides. If the engine is hooked up close to center, this is most likely the cause, as the movement from the eccentric and link are minimized.
- The other possibility is that the foot/basket of the link is not properly set which will cause the link to swing unevenly causing uneven valve movements. This will be minimized as the engine is hooked up closer to center.
- If one valve has more movement than the other, the throw of the eccentrics might not be the same or the link blocks are not in the same position in the link as each other. In forward, if one link block is higher than the other, it will have less valve travel than the other side.

Figure 63. When the piston is centered, the valve should be centered and the lead and lap lever should be vertical. With the lead and lap lever vertical, the valve will have even movement from both sides of being centered.

Setting the valve can be time consuming and frustrating, but it can also be very rewarding. If care is taken in measuring the valve movements and adjustments, a good running engine will be the end result (assuming the valve and the passageways are properly machined).

Setting the Eccentric Crank (Walscheart and Baker Valve Gear) Using a Tram Bar

The eccentric crank is a crank that is connected to the crank pin and has an offset pin to create another motion for the valve gear. The motion from the eccentric crank combines with the motion from the piston (through the lead and lap lever) to move the valve to let steam into and out of the piston chamber. If the eccentric crank is not set properly, the valve events will be off and the engine will be lame or out of square.

When setting the valves on the 2-4-0, 2-8-2 and 4-6-0, I did not use the tram bar method as I didn't really understand it. I set it by sight so the link was close to the same spot on dead center and when the engine was set on dead center, the valve didn't move when the valve gear was moved from full forward to full reverse.

When the eccentric crank is set, the link will be in the same position whether the engine is on front or rear dead center. (The length of the eccentric rod is adjusted after the link is set and the rod can be off and it does not affect the setting of the eccentric crank.) To set the eccentric crank, a tram rod (a rod that has points on both ends and is cut or bent to suit each engine and each specific place) is needed and a good, solid anchor point is needed (brackets, frame, cylinder) on the engine. (An anchor point that isn't solid but can move will allow the tram bar to move and produce poor results.) The tram bar should reach the foot of the link.

To set the eccentric, the engine is set on rear dead center and the tram bar scribes a line on the foot of the link. The

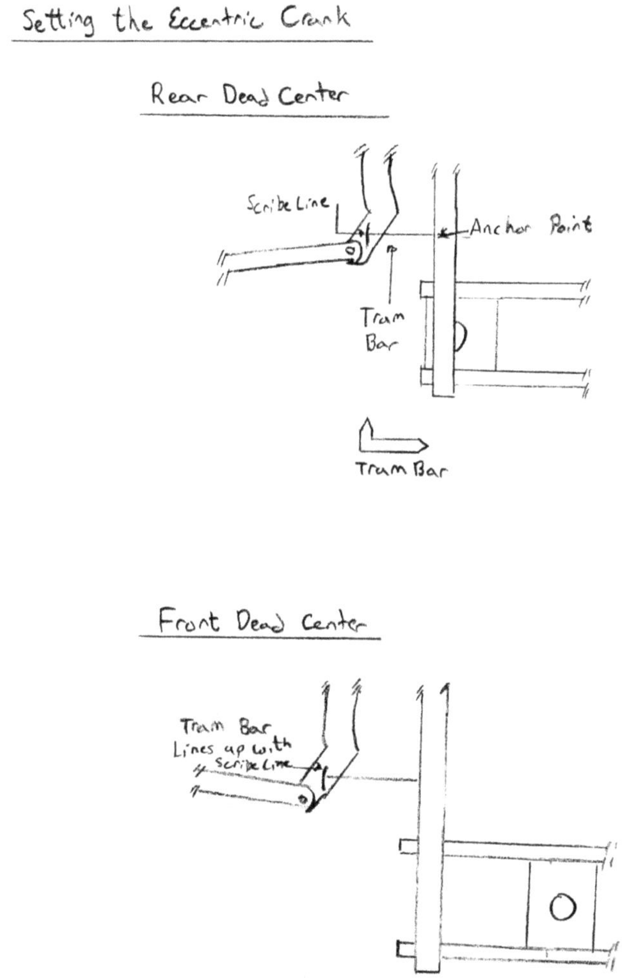

Drawing 15. When the eccentric crank is properly set, the tram bar at the foot of the link will scribe the same line when the engine is on front dead center or on rear dead center.

engine is then set on front dead center and the tram bar scribes a line on the foot of the link. If the scribed line at front dead center is at the same spot as at rear dead center, the eccentric crank is set. If there are two different scribed lines, the eccentric crank needs to be moved so the tram bar scribes a third line between first front and rear dead center lines. Once the tram lines up with the same scribed line at front and rear dead center, the eccentric crank is set.

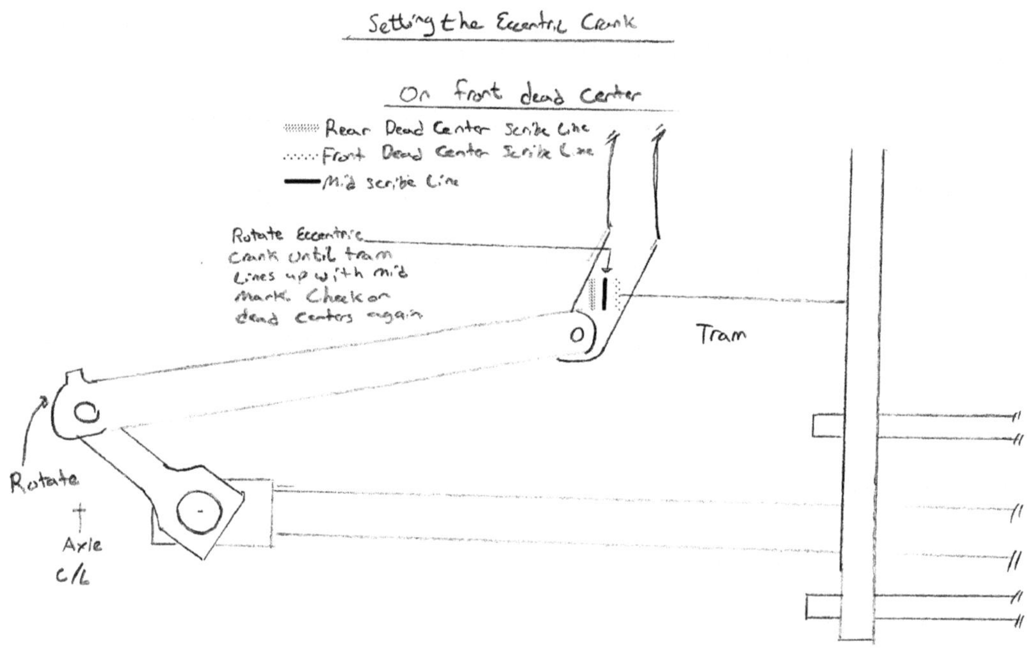

Drawing 16. If the tram bar scribes two lines on the foot of the link, the eccentric crank should be moved so the tram bar scribes a line between the two lines from front and rear dead center. In this sketch, the eccentric crank needs to rotate clockwise to properly set the eccentric crank.

Setting Eccentric Rods (Walschaert and Baker Valve Gear) Using a Tram Bar

When setting Walschaert and Baker valve gear, the valve is adjusted to get the proper lead when the valve gear is in center, then the eccentric crank is set and, lastly, the eccentric rod is set so the lead in full gear (and when everything is correct across the entire link) is the same as the lead when the engine is in center.

With the gear in full forward, the engine is rotated to one of the dead centers (drawing 17 shows the engine on front dead center). The valve is looked at through the sight hole to see its placement; is the lead correct (what it was with the gear in center), is the lead too big or is it too small? (Drawing 17 shows the lead is too small.)

A tram bar is used to scribe a line from an anchor/fixed point on the engine to the eccentric rod. (I used a point on the support for the crosshead guide supports as the anchor point.)

Then the engine is moved (in the drawing, the engine is rolled forward. If the lead is too big, don't roll the engine backwards but roll it forwards almost a full revolution, to keep the slop or clearance in the bushings behaving the same, until the desired lead is obtained) until the desired lead is obtained. The tram bar is then used to scribe another line on the eccentric rod. The distance between the scribe lines is how much the eccentric rod needs to be lengthened or shortened. (In drawing 17 the eccentric rod needs to be lengthened.)

If the eccentric rods have not yet been made, an adjustable eccentric rod can be made for setting the valves. It can be made out of ¼" × ½" steel bar stock or something similar. One half will have a hole with a few thousandths clearance for the eccentric crank pin at one end and two threaded holes (10–32 will work well) on the other end. The second half will have a slot (3⁄16" mill will create a good slot for 10–32's) to allow the two halves to be bolted together and be adjustable, and the other end will have a hole a few thousandths larger than the pin for the link foot. When the two halves are bolted together, set the length close to what it should be and put the adjustable eccentric rod on the engine.

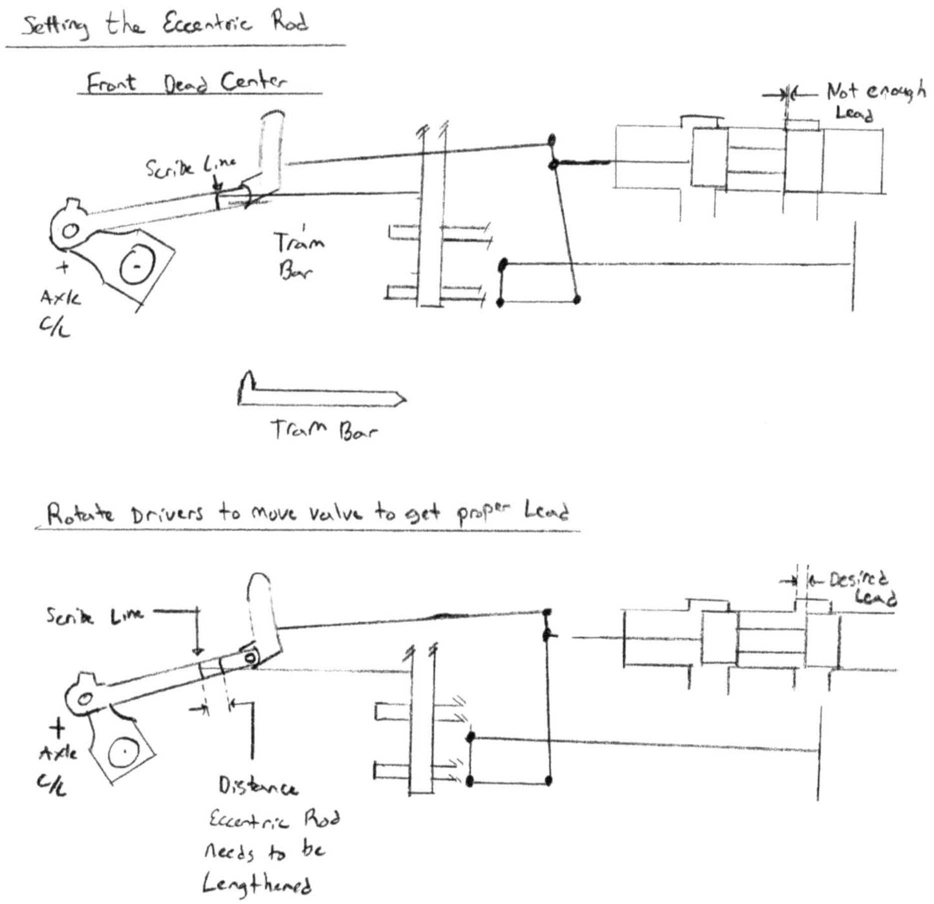

Drawing 17. The eccentric rod is set by placing the valve gear in full forward, rotating the driver to a dead center and looking at the lead. If the lead is the same as it was in center, the eccentric rod length is correct. If it is not, scribe a line on the eccentric rod when the engine is on dead center and then move the engine to obtain the desired valve opening. Scribe another line on the eccentric rod. The distance between the two lines is how much the rod needs to be shortened or lengthened.

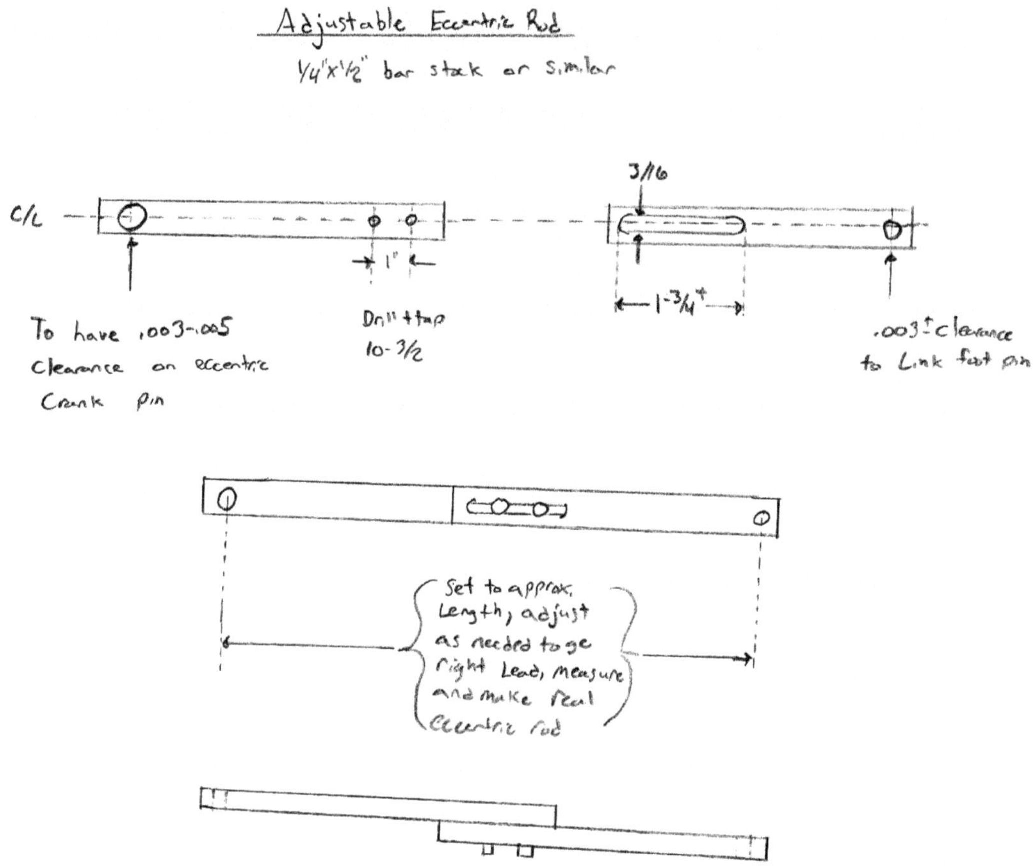

Drawing 18. A sliding, or adjustable, eccentric rod can be made and used to determine the needed length. The distance between the holes can be measured and the real eccentric rod made to those dimensions.

With the adjustable eccentric rod in place, rotate the wheels to the dead centers and adjust the rod length as needed to get the proper lead. Once the rod is set, take the rod off, measure the centerline distances between the end holes and make the eccentric rod to that length.

Adjusting the Eccentric Rod

On the full-size, the rods are adjusted by blacksmithing. The rod is measured cold, a section is heated up red hot and the rod is measured hot and then the rod is lengthened or shortened the desired amount. When the rod cools down, it will be the proper length.

On some O&K (German) engines, the eccentric pin bushing was bored eccentric and pinned in place. To make a length adjustment, the bolt holding the bushing in place was loosened and the bushing was rotated to lengthen or shorten the rod and the bushing was then held in place again with the bolt.

On miniature engines the rods can be remade to the desired length, one of the end holes can be filled in (weld, silver soldered) and then relocated, drilled or reamed at the

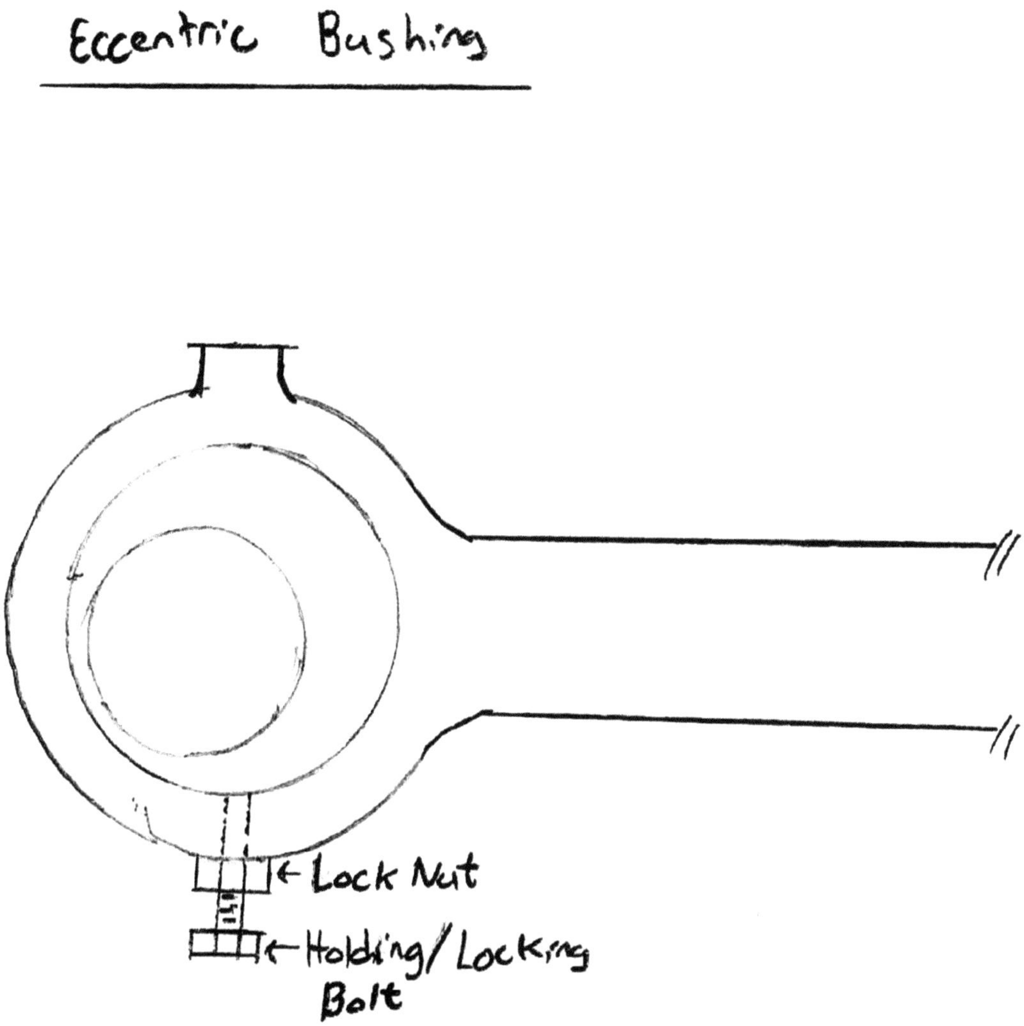

Drawing 19. Another way to change the length of the rod is to use an eccentric bushing for the crank pin bushing. As the bushing is rotated in the eccentric rod, the length between the holes will change. The bushing is pinned in place to keep it from rotating once the length is set.

proper length or the rod can be cut and material added or subtracted from the rod and then the rod can be welded or silver soldered back together.

Inside versus Outside Admission versus Indirect Motion

Inside admission means the steam from the boiler enters into the piston by passing by the inside edge of the valve (like the center of the spool valve). Outside admission means steam enters the piston by passing by the outside edge of the valve (like a Slide/D-Valve or a spool valve where the exhaust is in the center of the valve and steam from the boiler is on the outside edges of the valve).

With direct motion Walschaert valve gear for inside admission, the eccentric crank follows the crank pin and the radius rod attaches to the lead and lap lever above the connection to the valve crosshead. When the link block is in neutral, the movement of the valve is from the crosshead (through the lead and lap lever) and the valve movement is a parallel movement to the piston.

Figure 64. The 2-8-2 has inside admission valves with direct motion Walschaert valve gear.

With direct motion Walschaert valve gear for outside admission, the eccentric crank leads the crank pin and the radius rod attaches to the lead and lap lever below the connection to the valve crosshead. When the link block is in neutral, the movement of the valve is from the crosshead (through the lead and lap lever) and the valve movement is opposite from the movement of the piston.

Figure 65. The 4-6-0 has outside admission valves with direct motion Walschaert valve gear.

With indirect motion, instead of having forward being in the lower part of the link (below center) where the motion of the valve is the direct motion from the eccentric rod, forward is in the top part of the link where the movement of the radius rod is opposite the movement imparted by the eccentric rod. For inside admission Walschaert this means that the eccentric will lead the crank pin. Figure 66 shows a locomotive with indirect Walschaert valve gear with inside admission.

Figure 66. This former Pennsylvania Railroad 0-6-0 has inside admission valves with indirect motion Walschaert valve gear.

As a side note, for Baker valve gear, there are changes in the valve gear between inside and outside admission so for inside admission it is a direct motion valve gear but for outside admission it is an indirect motion valve gear.

Steam Blowing through the Piston and Valves

The piston rings on the piston and spool valve and the mating surface between the slide valve and the valve seat are supposed to create a steam tight seal, but sometimes they don't.

When we built the 2-8-2 Dad made piston rings. When the throttle was first opened it sounded like a jet taking off; great draft, not much power. Dad talked with his father (who was in charge of production at Hastings Piston Ring Manufacturing in Hastings, Michigan) and in his first sentence he listed at least four things that were not done properly. We bought properly or professionally made piston rings and that took care of the problem.

If the rings appear to be intact they can still leak because they don't have enough pressure against the wall of the chamber. This appeared to be the problem with the 2-4-0 when I got in her in Pennsylvania in 2012. She had a lot of blow by and even though the spool valve rings appeared to be decent, we replaced them and the blow by was greatly minimized.

Sometimes it is a single ring that is not sealing properly or is cracked or broken. Sometimes the bore is still pretty rough and the ring can't get a good seal or the bore is steam cut, which would need to be machined or honed out. Sometimes the valve or piston liners are not sealed in the cylinder block which will allow steam to blow by and escape out the stack.

If the engine is blowing more on one chuff than others, that could easily mean that when the valve is uncovering that exhaust port that steam is leaking by the piston rings on that end of the valve due to a broken or cracked ring that is not sealing (with spool valves) or that area of the valve chamber is steam cut allowing steam to blow by. With a slide valve, that end of the valve or valve seat may be worn enough so the valve does not have full contact with the seat or it is steam cut and needs to be machined/lapped back to being flat. (Generally speaking, if the piston rings are bad, the blow will happen on both strokes and not just one.)

If an engine has more of a constant blow the problem may be that the rings are bad on the piston(s) or valve(s) or that steam is leaking around the steam chest sleeve and leaking into the end exhaust ports or the valve seat and slide valve have worn enough that they cannot create a seal or there is a steam cut in the valve chamber or piston chamber.

Finding the source of the blow by can be a pain. The best I can suggest is start by taking out the pistons and valves and inspecting the bores and replacing the rings for starters and clean up any score or machine marks by gentle lapping or honing.

Otto Gas Engine Works (http://www.ringspacers.com/) is a supplier we have used for some of our piston rings.

Uneven Exhausts (Out of Square)

When an engine is considered to be "square" all the exhausts are of even strength/volume and they are spaced evenly apart. With most engines, there are four even exhausts per revolution of the drivers. Sometimes an engine is out of time/out of square with uneven exhausts. Parts may have worn, have shifted or been poorly designed or machined. This can result in an engine that is out of square. Here is a list of potential reasons why an engine may be out of square.

- Valve gear is worn and the accumulated slop does not allow for accurate movement of the valve
- One of the valve gear components has been hit, bent or moved out of position
- The piston/valve rings are worn, broken or cracked and are not sealing properly
- The valve and/or valve seat are worn
- Piston rings are stuck in their grooves
- The machined ports in the valve are not to the proper dimensions that match the valve

Specific to Walschaert valve gear:

- If the engine is on dead center and the valve moves forward when the Johnson bar moves forward and moves back when the Johnson bar moves to reverse, and this happens at front and rear dead center, then the eccentric rod is too long and needs to be shortened.

- If the engine is on dead center and the valve moves back when the Johnson bar moves forward and moves forward when the Johnson bar moves to reverse, and this happens at front and rear dead center, then the eccentric rod is too short and needs to be lengthened.
- If the engine is on one dead center and the valve is too far forward in forward and too far back in reverse, but on the opposite dead center the valve is too far back in forward and too far forward in reverse, then the eccentric is not set correctly. Once the valve is consistently too far one direction in forward and too far the other direction in reverse, then the eccentric is set, but the eccentric rod is not.
- If the valve is set correctly in center, but moves forward when the Johnson bar is put into forwards *and* moves forward when the Johnson bar is put into reverse, then the radius of the link is too small for the length of the radius rod.
- If the valve is set correctly in center, but moves back when the Johnson bar is put into forwards *and* moves back when the Johnson bar is put into reverse, then the radius of the link is too large for the length of the radius rod.
- If the cut-off is quite a bit uneven from the front side of the piston to the rear side, the problem may be in the position of the lead and lap lever and the length of the union link. When the valve gear is in center and the piston is centered in its stroke, the valve should also be centered in its stroke *and* the lead and lap lever should be vertical (set by the length of the union link) to allow even swinging from vertical and even valve movement forward and behind center.
- If one side has more or less lead than the other, the problem may be the positions of the holes on the radius rod (generally speaking, where the radius rod attaches and valve crosshead attaches) is not the same spacing from side to side which allows the different lead between the sides. If the engine is hooked up close to center, this is most likely the cause as the movement from the eccentric and link are minimized.
- The other possibility is that the foot/basket of the link is not properly set which will cause the link to swing unevenly causing uneven valve movements. This will be minimized as the engine is hooked up closer to center.
- If one valve has more movement than the other, the throw of the eccentrics might not be the same or the link blocks are not in the same position in the link as each other. In forward, if one link block is higher than the other, it will have less valve travel than the other side.

Specific to Stephenson valve gear:

- One or more eccentrics may have slipped or moved on the axles
- The rocker arm is not rocking evenly enough around a vertical plane
- The driving boxes have worn and the whole eccentrics/rods/links have moved forward/does not have a steady position in relationship to the cylinders
- The centerline where the lifting link connects to the link is not far enough behind the centerline of the link

In theory the valve gear should be able to give square exhausts no matter where the Johnson bar is placed. The reality is that there will be inconsistencies in the valve gear and timing as there will always be some slop in the parts as there needs to be enough clearance for the valve gear to move without binding. Even poorly designed valve gear can be set square in the "company notch" (when hooked up to the normal Johnson bar position when running forwards) as that is the notch that matters the most. Running in the corner is more forgiving to inconsistencies than when running hooked up.

Another Possible Reason for Uneven Exhausts

When setting the valves, the exhaust of the steam is just as important as the admission of steam. If the admission is dead even but the exhausting is not even, the exhaust out the stack won't be even.

Lap is how much the leading edge of the valve (which will admit steam to the cylinder) is covering the port to the cylinder when the valve is centered. If a valve has ⅛" lap, that means when the valve is centered an ⅛" overhangs the front port and an ⅛" overhangs the rear port (the valve needs to move ⅛" either side of center to become line in line with the port).

Lead is how much the valve is open when the piston is on dead center. If an engine has 1⁄16" lead, that means the valve is open by 1⁄16" (allowing steam to enter the cylinder) when the piston is on dead center.

If the exhaust side of the valve is also centered (but not overhanging the exhaust edge of the more than lap plus lead) the exhausting of the steam should also be even when the valve gear is set up correctly.

If one side of the valve is closer (or farther away) from the exhaust edge of the port than the other side of the valve, the exhausting of the steam will be uneven and will produce uneven chuffs even if the admission of steam is even.

If the exhaust side of the valve overhangs the exhaust side of the port by more than lap plus lead, then the steam won't exhaust before the steam is admitted to the other side of the piston. This can create a lame engine.

The error can be in a valve that is not machined properly or the valve seat or valve cage might not be machined to the proper dimensions or part of the port might be out of line with the other aspects which can allow the steam to exhaust (or be admitted) before it ought to be.

Drawing 20 illustrates these types of errors in the valve. They are not done to scale and should not be used to determine how the valve needs to line up with the ports. It would be rare for a valve to not have any lap with the exhaust ports, but I exaggerated the types of errors so they are more readily seen.

Adjusting the Valve Gear on the 4-6-0

Over the years I have revisited the Walschaert valve gear that I designed for the 4-6-0 to work out any bugs and carry out small corrections to the valve gear. As time becomes available, I have been going over the engine with a fine-tooth comb and tweaking things.

When running the 4-6-0 hooked up to the notch just forward of center (there are

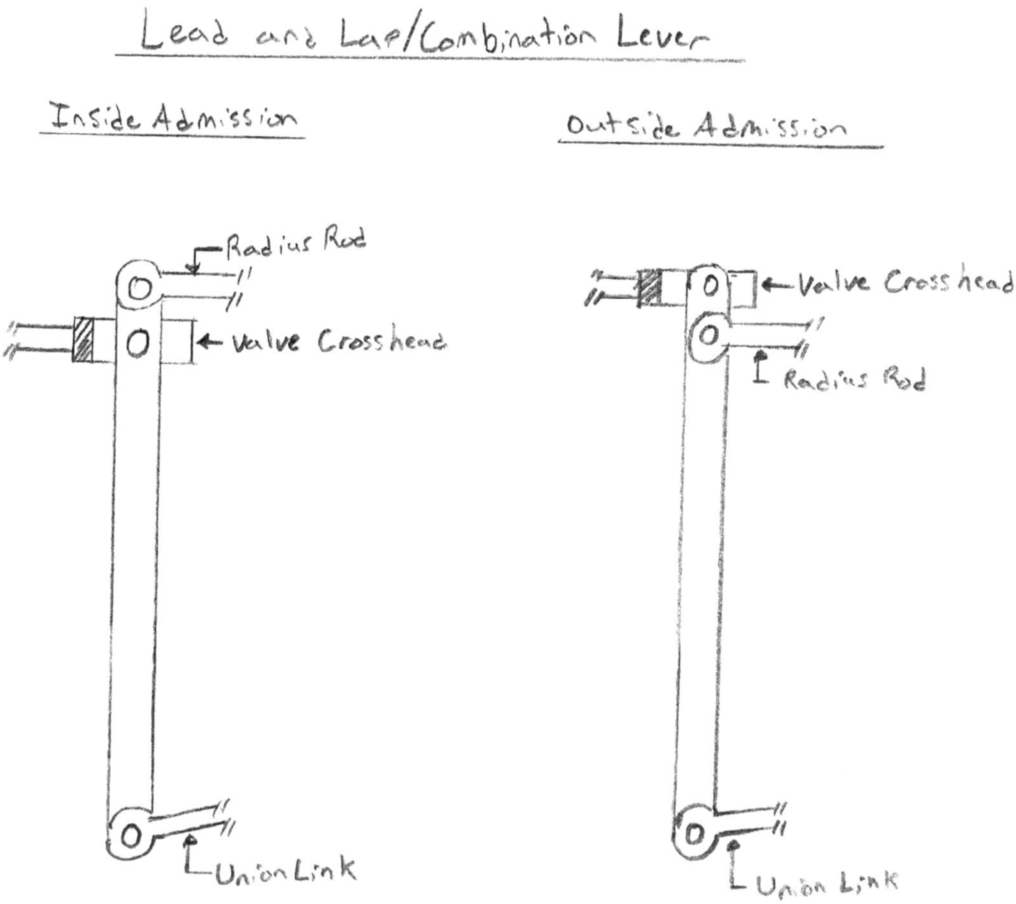

Drawing 20. Errors in the valve or the valve seats/cages can lead to uneven admittance and exhausting of steam to and from the pistons.

six notches in forward) she would pulse or surge when pulling a heavy load up a 3 percent grade. When she was pulling a lighter load, she would not surge and the exhausts sounded pretty even. I took out the inspection ports in the valve chests and the ports all had pretty even openings (about 0.050" in the notch just forward of center, and a strong ¼" in the corner) and the leads were all about the same. However, the cut-off was drastically different between the front and rear ports with the front port having a 32 percent cut-off and the rear port having a 64 percent cut-off.

With being hooked up that close to center, most of the valve movement was from the crosshead through the lead and lap lever. I began to wonder if the swing of the lead and lap lever was not centered on a vertical line.

I set the piston at its mid-stroke and the lead and lap lever was not even close to being vertical but the bottom of the lead and lap lever was too far forward. I measured that the union link was about 5⁄16" too long. I am not sure when I designed the valve gear in the spring of 2001 if I set the union link too long or if the main rods ended up being longer than I thought, therefore moving the crosshead too far forward.

I removed the union link and because I don't have a mill to use to locate the center of the holes, I did a layout using scribed lines and punched a mark where I wanted the new hole to be in the clevis end of the union link. I drilled and taped the new hole in the drill press and they came out within 0.005" of what I wanted. I reinstalled the union link (which had two holes in the clevis end now) and set the piston at its mid-stroke and the lead and lap lever was pretty much vertical now.

After I reset the valve I checked the cut-off and both the front and rear ports had about a 53 percent cut-off in the notch just forward of center. (The lap on the valve is 0.080" as that is how the castings cleaned up. If the lap were more, the cut-off would be less.) After the trial fit, I trimmed the excess length off the union link and reassembled the valve gear.

When I ran her, the surging was less when I had her hooked up to the notch just forward of center, but the exhaust was not quite totally even; it is close and when the engine is hooked down, there is not much if any difference between exhausts. The next project will be to take out and measure the valves and see how they compare to the ports on the valve seat.

Left: **Figure 67. With the piston centered, the lead and lap lever is not vertical due to the union link being too long, causing uneven valve cut-off.** ***Right:*** **Figure 68. With the piston centered and the hole redrilled in the union link, the lead and lap lever is vertical.**

Building Walschaert Valve Gear for the 4-6-0

Figure 69. The union link was trimmed. With the lead and lap lever vertical when the piston is centered, the cut-off is pretty even.

When Dad and I were building the 4-6-0 we decided that we would rather have Walschaert valve gear on her than the Stephenson gear that she had in the original design. We had already built two engines with the Walschaert gear and had a decent understanding of how it worked and liked the fact that this gear was on the outside of the frame where it could be worked on easily. Another advantage to Walschaert is that the link is rigidly attached to the locomotive frame and as the driving boxes wear and axles move forward, the eccentric crank and its rod will move forward, but because the link is attached to the frame, the error from the axles moving forward is minimized in the valve events. With the Stephenson gear being directly attached to the axle, as the axle moves forward, the gear also moves forward and the error is then directly imparted on the valve events. Since I no longer have my original sketches done on scrap paper, and I have had to make different adjustments as things have changed, I don't have any dimensions for the parts, but this is the process I went through to design the Walschaert valve gear. We built the engine over nine months during my junior year in high school, September 2000–May 2001.

While we were building the engine, I kept an eye out for where I could attach the brackets that would hold the supports for the links and tumbling shaft. I wanted to have the link a fair distance away from the main driver so that the angulation of the eccentric crank rod to the link would be minimized.

The placements of the supports are just in front of and just behind the first driver and the center of the link is just behind the centerline of the lead driving axle.

The eccentric crank is ⅜" thick with a ½" hole where it fits over the crank pin so I turned a ½" diameter by ⅜" long end on the end of the main crank pin. For ease of making the parts work, I decided to have the centerline of the eccentric crank be on the same centerline as the link.

Eccentric Crank

I wanted the eccentric crank to impart a ⅝" travel to the valve, but since the eccentric crank rod connects below full forward, the radius rod will move less than the movement

Top: Figure 70. The brackets holding the valve gear to the frame are in front of and behind the lead driving axle. These holes were located on the frame while the frame was being made in preparation for mounting the valve gear. *Left:* Figure 71. The eccentric crank is made out of ⅜" plate and the eccentric crank pin centerline is in line with the link centerline. *Above:* Figure 72. The eccentric rod body was made out of ½" square steel with the clevis end for the link made out of a piece of ½" × ¾" steel stock. The eccentric crank end was made out of ¾" wide stock, welded to the rod body and milled to width.

from the eccentric crank. If I remember correctly, the corner is two-thirds the distance from the center of the link as the eccentric crank rod connection is to the center of the link. This means to get ⅝" travel on the radius rod, the eccentric crank needs to move in a .938" diameter circle, or .469" throw. The eccentric crank is set 90 degrees off of the crankpin. To find out how long to make the eccentric crank I used the Pythagorean Theorem; $a^2 + b^2 = c.^2$ The length of the eccentric crank is then the square root of 1.625^2 (crank pin throw) + $.469^2$ (eccentric crank pin throw) = 1.691.

To set the eccentric crank, I drilled a 3⁄32" hole through the crank pin to use a 3⁄32" welding wire to help guide the eccentric crank to its 90-degree setting and measure the centerline of the eccentric crank to the center of the axle. For outside admission, the eccentric crank is set so it leads (in relationship to the crank pin when moving forward). If the crank pin is at the 6 o'clock position, the eccentric crank will be set at the 3 o'clock position.

Eccentric Rod

While I did calculate how long to make the eccentric rods and made them that length, I have needed to make adjustments to them since. Assuming that the radius rod length matches the radius machined in the link and that the eccentric is set properly, when the piston is set on front and rear dead center (meaning the wrist pin is on the exact centerline of the crank pins and all the force is transmitted horizontally and cannot rotate the wheel), when the Johnson bar is moved from full forward to full reverse, the valve should not move at all. When on the dead center, the valve should be open by the lead as determined by the lead and lap lever. I have had to make several adjustments to the eccentric rods on all of our engines (lengthening some, shortening others) to get the valve to be stationary when the Johnson bar is moved when the piston is at both dead centers. For setting the eccentric rod length, I would recommend making an adjustable eccentric rod to find the proper length and then make the real eccentric rod.

The link is also ⅜" thick. The radius rod is ¾" wide and has a ⅜" notch milled in it to fit around the link. The trunions are made out of 3⁄16" thick steel, milled to ⅛" thick where the bolts go through to accommodate the length of 5-40 socket head cap screws that we had. On the ends I welded a piece of .200" × ¼" steel to hold the trunions far enough away from the link so the radius rod would not rub on them. I set the internal width of the link supports far enough to allow 0.010–0.020" of lateral play on the link/trunion assembly.

Figure 73. The link was machined out of a piece of ⅜" thick plate. The trunions were made out of 3⁄16" thick plate with 0.200" thick shims welded to the top and bottom to step the trunion away from the link to make room for the radius rod.

Per the original plans, the centerline of the valve is offset ⅜" (if memory serves me correctly) to the inside of the centerline of the piston. The lead and lap lever (seen in figures 74 and 75) has an offset to go from the centerline of the union link to the centerline of the valve crosshead. I fabricated the lead and lap lever using ¼" × ¾" bar stock and angle iron. I bolted them together, welded them and then trimmed them to the width and shape that I wanted.

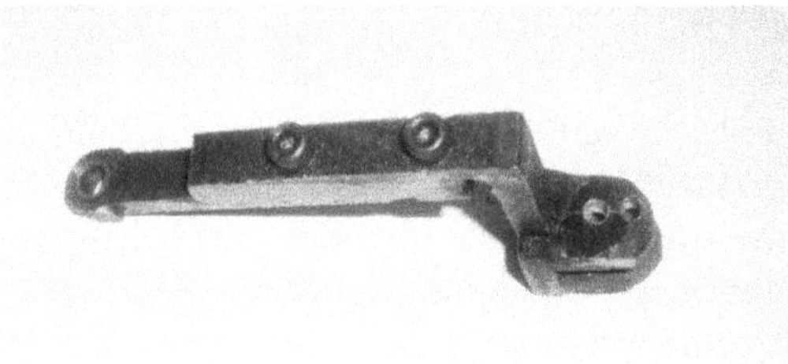

***Left:* Figure 74. The lead and lap lever is made out of ¼" thick steel and an angle iron in order to obtain the offset from the union link to the valve centerline. *Right:* Figure 75. The right lead and lap lever is mounted on the engine.**

On an outside admission valve (such as a D-valve) the radius rod connects to the lead and lap lever between the union link connection and the valve crosshead. When the engine is in center/neutral, the radius rod will act as a pivot to the lead and lap lever. As the piston moves forward, the valve will move backward, and so forth. The purpose of this is that Walschaert gear has a set lead (how much the valve is open on dead center) and lap (how much valve overhangs the steam inlet ports when the valve is centered) and the movement of the valve through the lead and lap lever is 2x (lead + lap).

Lead and Lap Lever

The lap on the 4-6-0 is 0.080" with the way the castings cleaned up and I wanted a lead around 0.020". I decided to make the lead and lap lever give around 0.045" lead (giving around 0.125" movement off of center, around .250" total movement) to account for clearance in the bearings that will give lost motion; if there is 0.005" clearance between the radius rod bushing and pin and 0.005" clearance between the valve cross head bushings and pin and 0.010" clearance between the valve steam and the valve, that gives a total of 0.020" of lost motion, reducing the lead from 0.045" to 0.025".

The piston travels 3.25" and I want the valve to move 0.25" so the lead and lap lever should give a lever ration of 12.93:1, meaning the valve moves 7.73 percent the distance of the piston. Using the diagram of the Walschaert valve gear for inside admission for a 4-6-2 (drawing has valve gear superimposed over two drivers) I found that the maximum movement of the lead and lap lever should be 50 degrees (25 degrees off of vertical) maximum per this diagram for this engine.

Using this, the dimensions of the lead and lap lever came out: 3.75" from the union link connection to the radius rod connection and 0.290" from the radius rod connection to the valve crosshead connection. This gave a valve movement of 0.252" and a total movement of the lead and lap lever of 51.3 degrees, 25.6 degrees off of vertical.

Valve Crosshead

I fabricated the valve crosshead so it would fit around the lead and lap lever and follow a round crosshead guide. I decided to use a ⅜" diameter O1 drill rod for the guide as I was not sure if I would be able to get a square guide aligned close enough so it would not bind. The guide was threaded on the lathe so the threads are straight and the guide is secured by threading it into a ⅜" × 1" plate that is bolted to the top of the valve chest. To fit up the guide, I machined the top of the valve cover flat and then machined the ⅜" × 1" plate until the guide was parallel to the valve rod.

***Left:* Figure 76. The valve crosshead body is made out of a piece of ⅜" thick steel and the sides are made out of ¼" × ¾" steel (bolted and welded to the body) and the guide is a piece of ⅜" O1 drill rod. A bronze bushing is used in the crosshead body for a bearing surface to the guide. *Right:* Figure 77. The union link is made out of a piece of ½" × ¾" steel milled down to ¼" wide on the back and made as a clevis to fit over the lead and lap lever on the front. Due to the length of the lead and lap lever, the mount for the union link to the crosshead (I used a piece of ¼" thick steel bolted to the crosshead) is below the wrist pin.**

Union Link

The union link is used to transmit the movement of the piston to the lead and lap lever. Per the diagram of the valve gear on a 4-6-2, the length of the union link is to be around ½ to ¾ the length of the stroke. In order to have an even movement of the valve in relationship to the piston, when the piston is centered in its bore, the valve should also be centered and the lead and lap lever should be vertical.

In order to minimize the angulation effect and to make it easier to make, I decided to connect the union link below the wrist pin on the crosshead, but in the wrist pin's vertical centerline. I bolted a ¼" thick plate of steel to the crosshead to use for the union link attachment.

Radius Rod

When the valve is centered and the lead and lap lever is vertical, the distance between the radius rod connection in the lead and lap lever to the center of the trunion bearings is the length of the radius rod and sets the radius to which the link will be machined. I fabricated the radius rod using ½" square stock for the body, ½" × ¾" stock for the end that goes around the link and angle to move the centerline from the center of the link to the center of the valve. I bolted the pieces together, welded them and then machined the radius rod.

Link

The link was machined on the radius as determined by the radius rod. If the length of the radius rod does not match the radius of the link, the error can negatively affect the timing and steam distribution. Per the Walschaert diagram, I kept the angle of the radius rod around 1" in 12" incline when running in the corner. The holes for the bolts for the trunions line up with the pattern in the trunions and leave ample clearance so the radius rod does not interfere with the trunion when running in the corner.

The foot of the link I set on the axle centerline. Per the diagram showing the link movement, I set the connection to the eccentric rod behind the centerline of neutral/center in the link slot to even up the movement of the link. I did not find any information as how far behind it should be, so I set it 0.75" behind as that is how it is set up on the 2-8-2 and she runs fine.

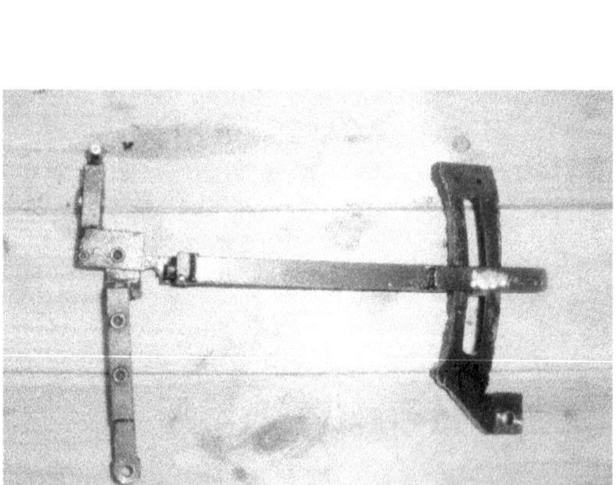

***Left:* Figure 78. The left lead and lap lever, radius rod, valve crosshead, link and link block are assembled. *Right:* Figure 79. The right link, radius rod, lead and lap lever, and valve crosshead are assembled on the engine.**

Tumbling Shaft and Lifting Links

Due to personal preference I decided to mount the tumbling shaft behind the link (like our other engines) and during the initial building I thought I had located it far enough behind the link so it would not interfere. During the assembly of the valve gear I found out I had to move it back about ½" more so it would not interfere. The lifting links were made so both sides of the engine would have the radius rod on center at the same time and are made out of by bending ⅛" × ½" steel stock.

Left: **Figure 80. The tumbling shaft is made out of ⅜" diameter bar stock and the arms are made out of rectangular bar stock and welded together for the reach rod connection to the Johnson bar.** ***Right:*** **Figure 81. Due to the clearance under the boiler, the tumbling shaft is straight. The pillow blocks had to be moved about ½" back from my original guess for their location.**

Adjustments

When the valve gear was first assembled and the engine ran on air, she ran decently but I needed to make some adjustments as what came out on paper did not perfectly match what came out in reality. As I have taken a closer look doing inspections I have had to make other adjustments. I have had to adjust the eccentric rods after new driving boxes were made as the main driver moved slightly compared to the valve gear. I had to shorten up the union link to even up the valve events. I am not sure if I missed that during the initial building or if the crosshead is farther ahead than where I thought it would be. I had to make an adjustment to the left lead and lap lever as the lead on the left side was about 0.015" less than the right side, even after replacing worn pins and bushings.

When running, we hardly ever run in the corner, unless we need full power. The engine does use less water when she is hooked up. She has six notches in forward. When pulling five empty cars around my 485' loop (which has a 3 percent grade), starting with a full tender of water and running 1–2 notches forward of center, I can go 24 times around the loop and there will be 1.5" of water left in the tender. If I run in the corner with a five-car empty train, I can run 18 trips before there is 1.5" of water left in the tender. When pulling a five-car train with five adults and two kids up around the track (needing full throttle and in the corner to go up the 3 percent grade) I can get 14 trips before the water is down to 1.5" left in the tender.

When the valve gear is set well, she runs well in all the notches; she has six notches in forward. In the corner the cut-off is around 92 percent with a port opening of ¼"+.

When the gear is set one notch forward of center, the valve is opening around 0.050" and has a 50 percent cut-off. If the lap were greater, the cut-off would be shorter, but due to the way the castings cleaned up, this is what we have.

I have been very satisfied with the Walschaert valve gear on the 4-6-0. I enjoyed designing it and it was a good way to use the trigonometry I had learned my last couple of years in high school.

Lead and Lap/Combination Lever

The lead and lap lever is used with the Baker and Walschaert valve gear. It is connected to the radius rod, the valve crosshead and the union link (which connects to the crosshead). The lead and lap lever is used to transmit a portion of the piston movement to the valve. When in center on Walschaert gear, the radius rod is a fixed point and all the motion in the valve comes from the piston movement. The radius rod becomes a pivot point and the lead and lap lever is a simple type of lever that moves the valve in a small proportion to the movement of the piston.

The lead and lap lever connections are placed differently between inside and outside admission valves. Inside admission valves (spool valve) admit steam from the inside edge/center of the valve. Outside admission (D-valves and some spool valves) admit steam from the outside edge of the valve.

With inside admission, the radius rod connects above the valve connection. With inside admission, the valve will move forward to uncover the forward port as the piston moves forward to front dead center. The valve motion is in the same direction as piston so the valve connection is on the same side of the pivot (radius rod) as the piston. Here is a video of the valve movement with the Walschaeret gear in center on an inside admission 2-8-2: https://www.youtube.com/watch?v=2eGy9IWGZm4.

With outside admission, the radius rod connects below the valve connection, in between the piston and valve connections. With outside admission, the valve needs to move backwards to uncover the front port as the piston moves forward to front dead center. The movement of the valve is opposite the movement of the piston so the pivot (radius rod) is placed between the valve and piston connections to reverse the movement. Here a video of the valve movement with the Walschaert gear in center on an outside admission 4-6-0: https://www.youtube.com/watch?v=pyJDDBPm3dk.

"Lead" is the term used for the amount that the valve is open when the piston is on dead center. "Lap" is how much the valve is covering the steam ports when the valve is centered. When the valve is centered, the lead and lap lever will move the valve ahead a distance equal to the lap plus lead. The lead and lap lever will also move the valve behind the center position by a distance equal to lap plus lead. The total movement of the valve from the lead and lap lever is 2 (lap + lead).

If the lap and lead are the same on engines with different length strokes, the dimensions of the spacing of the holes/connections on the lever will be different. If two 1:8 scale engines have ⅛" lap, 1⁄32" lead, but one engine has a 3" stroke and the other engine has a 4" stroke, the spacing between the connections will be different to have the valve move the same amount (2(lap + lead) which is 2(⅛+1⁄32) = 5⁄16" movement) on both engines. To make it easier, I kept the overall length (5.000") of the lead and lap lever the same and just changed the stroke length.

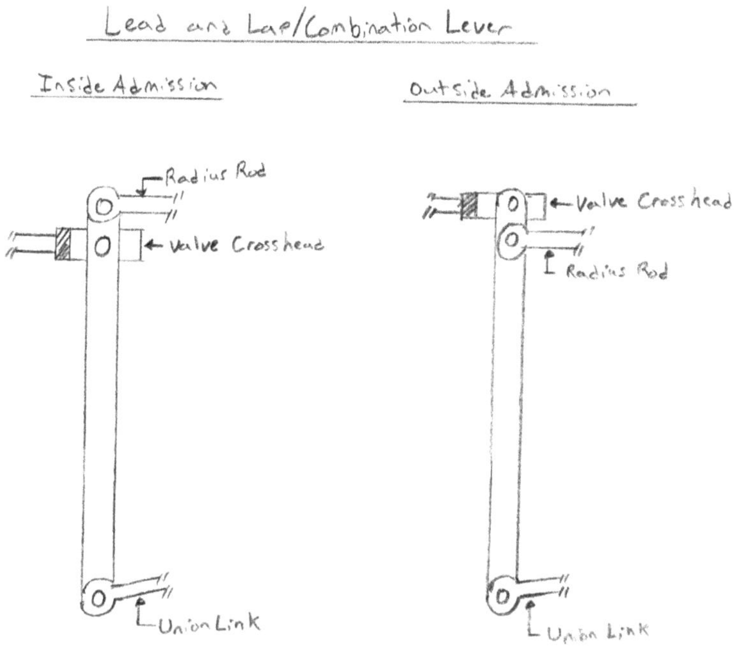

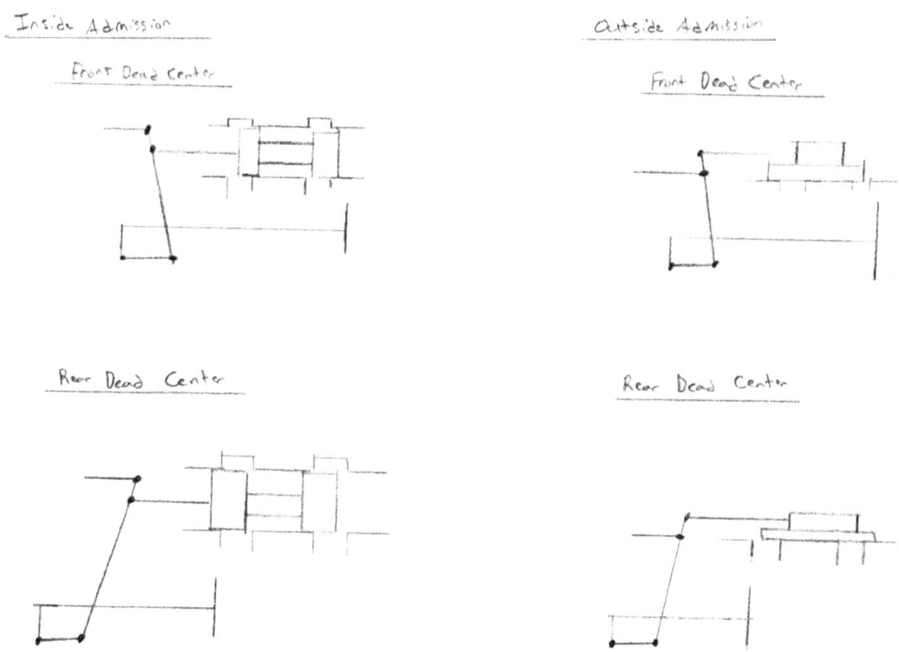

Top: Drawing 21. While the union link connects to the bottom of the lead and lap lever, the connections for the valve crosshead and radius rod are different depending upon if the engine has inside or outside admission valves. *Bottom:* Drawing 22. When the engine is on dead center, the valve will be open for the "lead." This is done by the movement of the piston through the lead and lap lever. With inside admission, the valve moves in the same direction of the piston but with outside admission the valve moves in the opposite direction. This depends upon the connections of the valve and the radius rod to the lead and lap lever.

With inside admission and a 3" stroke, the distance between the radius rod connection and valve is 0.520". For a 4" stroke, the radius rod to valve connection is 0.390".

With outside admission and a 3" stroke, the distance between the valve connection and the radius rod connection is 0.471". For a 4" stroke, the valve to radius rod connection is 0.362".

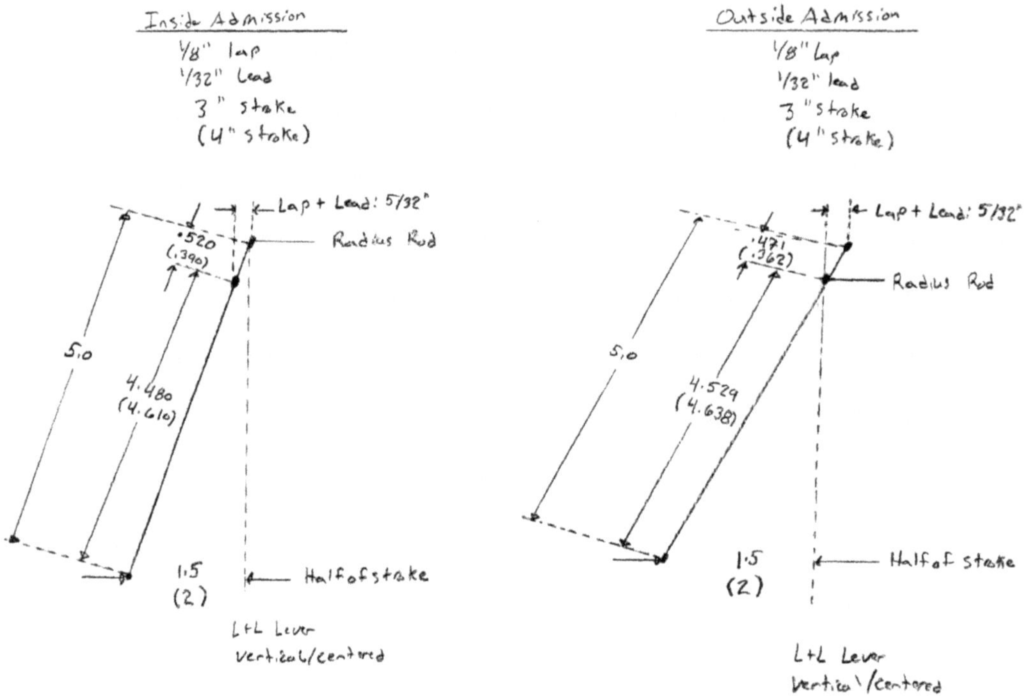

Drawing 23. Some sample dimensions are given for how to figure out the lengths of the lead and lap lever.

The lead and lap lever lengths are set up when the engine is designed and has specific lengths for each class or type of engine. Unless something major changes (increase/decrease lap, want to increase/decrease lead or the stroke changes) the lead and lap lever dimensions are set and don't need to be messed with.

Rebuilding the Valve Gear

When there is about 0.010"–0.012" slop between the pins and bushings in the valve gear, I will tear it apart and replace the worn bushings and pins. While 0.010" doesn't seem like much, it adds up in a hurry. If there is 0.010" slop between the link block and the link, link block and its pin, trunions and their bushings, between the radius rod and the pin connection to the lead and lap lever, the bushing in the lead and lap lever and the pin connection to the valve crosshead and between the valve crosshead and its pin, there is a total of 0.060" of accumulated slop at the valve, and that does not include any slop with the union link or eccentric rod.

A slop of 1⁄16" is a lot more in miniature than full-size. If a full-size valve moves 6" and there is 1⁄16" of slop, the lost motion is 1 percent of the valve movement. However, if the valve travel is ¾", a 1⁄16" of slop is 8.3 percent of the valve's travel. This will greatly affect the accuracy of the valve events.

If the lead is 0.020" on a miniature engine, the lead will greatly be diminished or become non-existent if the slop becomes too great. If the slop is too great, the engine will start to sound lame when hooking the engine up and the steam distribution won't be nearly as even.

For the most part the pins are made from socket head cap screws (cut them to length and they are harder than mild steel) and I use sintered bronze bushings. I buy new bushings, press out the old ones, press in the new ones, drill an oil hole and debur the edges.

For the link blocks, since I don't have a mill, I take a piece of bronze that is the correct width, drill and ream the pivot hole in the bronze, coat it with dykem blue and then lay the link on top of the bronze. I use a scribe and trace the curve of the link onto the bronze. I cut the link large and then use a grinder to remove most of the excess material and then file the link block to shape and keep checking it against the link. With care the link block will have a good, nice and smooth sliding fit without extra slop.

A big part of keeping the miniature steamers running well is keeping the valve gear in good condition. If an engine is being run a lot, the valve gear will wear out faster than if it's only being run a couple of times a year. The frequency of rebuilding depends on the owner and how often the engine runs.

Systems, Improvements and Why Things Are Done in Certain Ways

After a steam locomotive is built or rebuilt, the engine will need to be broken in and parts inspected to look for signs of trouble (inspections are covered in the Appendix) and things that need to be adjusted. Problems could include, but are not limited to, poor draft, poor steaming ability, engine too stiff/binding, valve gear out of time, leaks at the cylinder heads/steam chests, leaks in the piping, and more. Sometimes it may take some time for problems to present themselves.

As any equipment (steam locomotives, tenders, cars, and diesels) runs, parts will start to wear out. Bearings will become sloppy and need to be replaced, pins will wear out and need to be replaced, wheels and flanges will wear and will need to be replaced or have tires sweated on, driving and journal boxes will slowly wear and become sloppy in their fits over time and truck bolsters can wear and have a poor fit to the car frame. Taking care of these items is an important part of maintenance to address normal wear and tear.

In addition to these, many builders like to add items (steam pumps, injectors, generator, sanders, tender brakes or other detail items) or make improvements to their locomotives over time. This section addresses multiple different things that fall into these three aspects.

Potential Reasons for a Poor Steaming Engine

I have been asked many times why an engine is a poor steamer. Here are some things I would look at if an engine is not steaming well.

Air leaks in the smokebox. If the smokebox door does not seal tightly on the smokebox or the smokebox does not seal tightly on the front of the boiler or there are any holes that allow air to enter into the smokebox, air from outside the smokebox will enter into the smokebox through these leaks instead of coming up through the fire. The blower and blast nozzle may be properly set, but the partial vacuum they create will draw air in through the leaks instead of drawing it through the fire.

Steam leaks in the smokebox. Leaks from steam lines (blower, lines to cylinders, blast pipe/nozzle) or from tubes or tube sheet will fill or minimize the partial vacuum that is created in the smokebox so less air will be drawn through the fire.

Leaks in the firebox. Leaks from flues, staybolts, sheets or seams can fill the firebox with steam which will minimize the amount of air that can be drawn in through the fire as it will partly fill or minimize the partial vacuum that is created in the smokebox. If a leak is bad enough, water can put out the fire or severely dampen the fire.

Blast nozzle not the right size. If the blast nozzle is not the right size, the blast may not fill the stack completely to create the best draft or may be too strong or too weak. It seems for coal burners the best blast nozzle diameter (single port blast nozzle) is 14–18 percent of the piston diameter and for oil 18–22 percent of the piston diameter.

Blast nozzle not centered in stack. If the blast nozzle is not centered in the stack, the exhaust may not fill the entire stack, which can create poor drafting

Not enough space between the blast nozzle and the petticoat. If the blast nozzle is too close to the petticoat, the draft can be severely minimized even if it is centered and is the proper size blast nozzle. I have found that 2–3" between the blast nozzle and the petticoat works well on 1:8 steamers.

Boiler is scaled up. If the boiler is full of rust and scale, the heat transfer from the fire through the sheets to the water can be greatly reduced.

Tubes are dirty. If the tubes are dirty the heat transfer can be greatly reduced, and the volume that can pass through them is reduced. If I remember correctly, 1⁄32" of carbon has the same insulating quality as 3⁄8" of lime scale. If carbon is left in the tubes, over time and with use the carbon can become hard and will really take a scrubbing to remove.

Not enough air entering the fire. If the fire is not getting enough air, steaming qualities can be reduced. For a coal burner the space between the grate bars can be increased (on the 2-4-0 the grate bars are 3⁄16" wide with 5⁄16" space between them. The 2-8-2 and 4-6-0 have ¼" thick bars with 5⁄16" space between them). On an oil burner, air openings can be added at the back of the oil pan (not at the front as the air will take a shortcut to the flues, not be heated up, and it can thermal shock the flues).

Too much air is entering the fire. I have seen this more on oil burners or propane than with coal. If too much air enters the fire not all of it will be heated so cool air will be heading through the tubes. This will cut down on the steaming qualities and thermal shock the firebox and flues and can shorten their life.

Boiler is not designed properly. I have seen this on engines built in the 1930s, 1940s and 1950s that were not built from kits, and in designs developed by suppliers to the hobby. These engines are built and designed to the best ability of the builder, which at times can lead to designs that are less than ideal. The boiler may be too small for the demand that is placed on it, or the firebox is too small, among other design problems.

Stack is not designed properly. This is one that I have seen with engines that are not built from kits or in designs that have been developed by suppliers to the hobby. One engine I saw had a stack that was too small in diameter (it also looked like it was too short) so the volume of gases that needed to pass through the stack could not do so.

Poor timing. If an engine is poorly timed (too much slop, poorly designed, parts are not machined right, parts have moved, etc.), the distribution of the steam won't be right and the engine can waste steam and use a lot of steam due to poor timing.

Helping an Engine Become a Better Steamer

With the 2-4-0 being a small boiler (running level is about two gallons), it can be pretty unforgiving at times if you get too far behind with coal and water. She is also not as forgiving as the 2-8-2 (running level is about eight gallons) and when we started she seemed to get short of breath quickly. Here are things that we did to help the steaming capacity and reduce the usage of steam. We have done these things to the 2-8-2, 4-6-0 and the 15" gauge 4-4-2 and it all helped their steaming abilities and made them easier to run. We increased the air space in the grates, installed an arch (this made a huge difference), got rid of the axle pump and now use injectors and better coal and we hook the engine up.

For coal. We increased the amount of air flow through the grates. The 2-4-0 grate bars were ⅛" thick with ⅛" space between them. Her new grates have bars are 3⁄16" thick with 5⁄16" space between them. The other engines have ¼" thick bars with 5⁄16" air space.

Coal. Different grades of coal burn differently. Some produce a lot of heat with a lot of soot, others produce a lot of heat without a lot of soot, others are ashy, etc. If the grade of coal is not good it can be difficult to get the boiler to steam freely.

Firing technique. Different engines and coals can require different firing techniques. With the smaller fireboxes on 1:8 and smaller engines, the coal likes to solidify. The 2-4-0 and 2-8-2 like to have the fire raked a medium amount, but the 4-6-0, with only a 4" wide grate, really likes to have the fire raked a lot more, especially in the back. But on a 15" gauge 4-4-0 we don't rake the fire as it burns anthracite and raking would stir up the ash and cause the anthracite to clinker.

For oil. We made new burners that produced a bright yellow flame, even when turned up. We used as low a steam setting on the burner as possible (the more steam used, the less volume in the firebox for air) and made a new oil pan to allow minimal air to enter by the burner (will take the shortcut and jump right into the bottom flues without being heated up) and the main air supply comes in from the rear of the pan, controlled by a damper. Too much air flow can hurt though and have a cooling effect.

Cleaning the flues. We used a 50-caliber brush to clean the 2-4-0's and 4-6-0's flues (½" ID) and a 10-gauge brush to clean the 2-8-2's flues (¾" ID). The brushes are bronze and do a good job of cleaning out the carbon build-up. The brush needs to be strong enough to clean out the build-up (nylon is too soft) as carbon is a great insulator and will cut down how much heat can be transferred through the tubs.

Blast nozzle. For a single round port blast nozzle, the research I have done shows that the blast nozzle for coal should be about 14–18 percent of the piston diameter and for oil should be about 18–22 percent of the piston diameter.

Blast nozzle/petticoat placement. The best results I have had are when the distance between the blast nozzle and petticoat is ¼–⅓ the smokebox diameter (blast nozzle should be centered in stack).

Boiler washes. Doing periodic washes of the boiler to remove scale helps the boiler to steam better by washing away the scale that impedes the transfer of heat.

Firebox Arch

When my dad and I first started running our Railroad Supply 0-4-0, later modified in 2001 to a 2-4-0, in the spring of 1996 we had trouble keeping up steam. The safety valve was set for 130 psi, but many times we would drop down to 60 psi after running for less than a mile. Once we learned how to fire the engine, we were able to keep the pressure up … for a while anyways.

All three of the engines we built are coal burners with our 10-wheeler able to convert to propane in about 45 minutes. The 0-4-0 was the first engine we (actually mainly my dad, since I was six when he started it) built and was our first coal burner. My dad had bought a little Shay that burned kerosene in a pot burner and that is how we first got started. It was slow and needed some work but it got us by until we completed the 0-4-0.

During the last Tri-State Locomotive Club meet at Bill and Paul Fitt's track, someone suggested that we might want to install an arch in the firebox to help with combustion and steaming ability. Figure 82 shows the brick arch that is installed on the Strasburg (former Canadian National) Rail Road #89.

Figure 82. Steam locomotives used brick arches (such as this one on former Canadian Nation #89, a 2-6-0) to increase the path of the gases and flames to give a longer burn time and cut back on the amount of unburned fuel escaping through the stack as black smoke.

The brick is supported by three arch tubes that run from the throat sheet below the tubes to the door sheet about the firebox door. Very few model steam engines have arch tubes so we had to figure out a way to support them in the firebox without having the support be destroyed by the heat and not have it ruin the steaming capacity. It was not until 2000 or 2001 that we figured out how to do it.

The arch is up against the throat sheet and extends out to the side sheets so only a minimal amount of air can pass by without being deflected by the arch. The arch is below the tubes so it protects the tubes from being exposed to the direct heat and flames of the fire thereby prolonging flue life. The arch will also help to heat up any air coming into the firebox through a hole in the fire and minimizing the temperature shock on the flues when a hole develops in the fire. Also, since the arch is exposed to the heat from the fire, it will also hold the heat when the firebox temperature drops and will help heat up the gases to get a better combustion.

When fresh coal is added to the fire, the temperature inside the firebox drops until the green coal is brought up to a temperature where it can burn. The more coal that is added, the longer it takes for the temperature to rise and the longer it will take for the boiler pressure to rise; this is why filling the box up is not as efficient as firing lighter and a little more often.

The hydrocarbon gases may start to be driven out of the coal around 400°F, with most of them being driven off by 900°F. However, the temperature needs to be around 1,800°F in order for the hydrocarbons to break up into free carbon and free hydrogen and burn. If the temperature is less than 1,800°F, the hydrocarbons pass through the flues (where the temperature drops quickly) and up the stack as unburned, wasted fuel.

The arch will not let the gases in the front of the firebox pass directly into the flues. The cooler gases will hit the hot arch (which acts as a heat bank) and the hot arch will help raise the temperature of the gases so they can burn as well as have the gases pass over the fire before heading through the flues. This helps to cut down on the smoke and wasted fuel.

The arch length is determined by how high the arch can go. The area between the top of the arch and the crownsheet needs to be 115 percent to 185 percent (depending on which document you reference) of the net flue area. For models that do not have arch tubes that determine the height, the arch usually extends ⅓–½ of the way back from the throat sheet.

The formula for finding the net flue area is:

R = radius of the tube (half the diameter)
$\pi = 3.1416$
Net Flue Area = $R^2 \times \pi \times$ number of flues
Area above Arch = $1.2 \times$ Net flue area
= Firebox width × height above arch

Applying this to the 1:8 scale engines proved to be a little more interesting. In 1996 I first started out with a frame made of mild steel to support the arch. The frame's legs rested on the grate supports so it took up a little bit of grate area. That did not help, and the frame was too cumbersome and took up too much room in the firebox and the steaming qualities were decreased. In addition to that, the mild steel cannot take the heat and would burn out very quickly. I used mild steel as we had it and this way I found out without buying stainless that this way of holding the arch would not work.

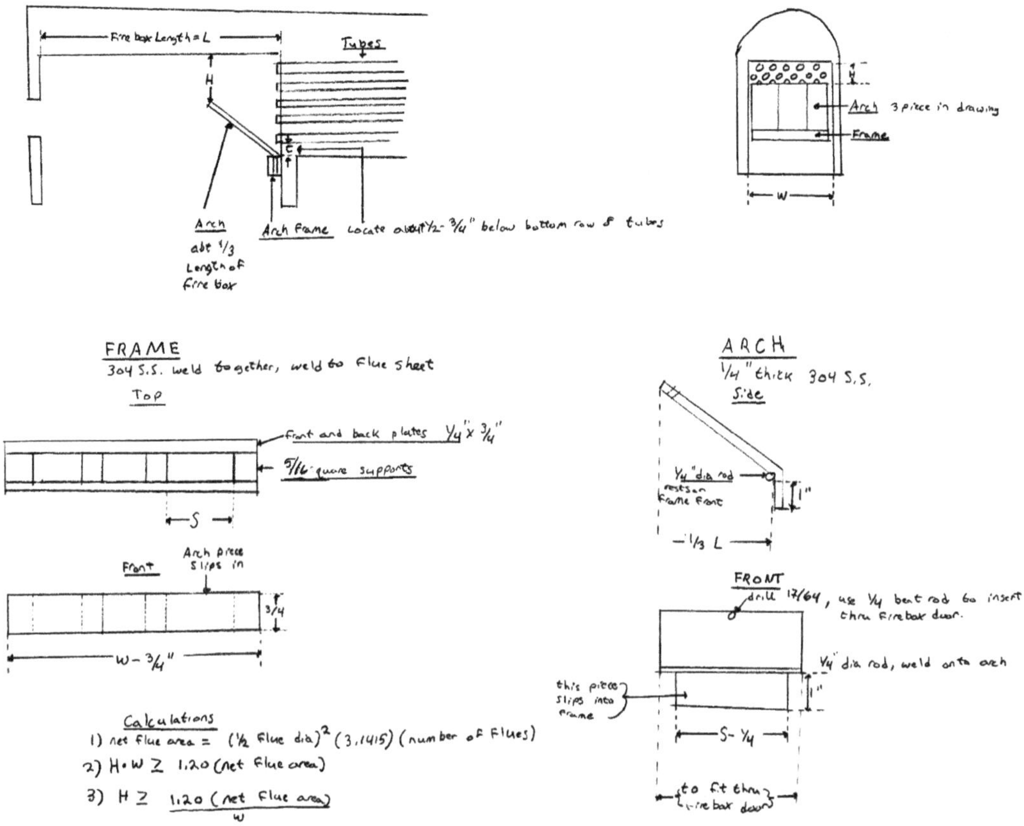

Drawing 24. This drawing shows how I installed the arch and how to determine how high the arch should be.

Around 2001 or so I thought about welding a holder/frame for the arch to the rear tube sheet. This would hold the arch 0.5"–0.75" below the first row of flues and would not affect the grate area nor would it reduce the firebox volume by much (unlike the frame I had tried before). The frame (see drawing 24) is made with 316 stainless steel (if 316 S.S. cannot be found, then 304 works well too) and is welded together so that the arch will slip into the slots. The number of slots and size depends on the size of your firebox. The height of the frame should be around ¾"; any shorter and it is easy for the arch to work out of the frame during use.

The pieces of the arch are made from ⅛" thick (for the 2-4-0; ¼" thick for the 2-8-2 and 4-6-0) 316/304 S.S. and have a "tail" that is 1" long that will set into the frame. The width of the arch is such that it can fit through your firebox door and the length of the arch is about one-third the length of the firebox.

For bending the arch and figuring out how high the arch should be at the end, you need to calculate your net flue area. Then multiply that by 1.2, and then divide by the width of the firebox. This will tell you how much height should be above your arch. If it is less than this distance, the fire can be choked off as not enough air will be able to pass around the arch.

The arch can also be made out of thinner metal if it is easier to find. We also found that we needed to weld a ¼" round bar to the arch at the bend on the rear/fire side of the arch to rest against the frame. This helped to prevent the arch from falling on the fire during use.

We welded the frame together outside of the firebox and when we welded the frame to the rear flue sheet, we filled the boiler with water to act as a heat sink to make sure we did not overheat the sheet or tubes.

The arches are easy to remove through the firebox door. We drilled a hole in the end of each piece so our poker stick, for the fire, would fit into it and be able to lift the arch in and out.

Figure 83. Due to the small size of the 2-4-0's firebox door, the arch was made in three pieces. Since the 2-8-2 and 4-6-0 have large doors, their arches were made in two pieces. The arch has worked very well with the 2-4-0 (sorry about the piece of coal on top of the arch).

In addition to cutting down on the build-up of carbon (before the arch we would have to brush out the flues on the 2-4-0 every two-and-a-half hours or so. With the arch we can run around five hours and the flues will be dirty, but not really limit the steaming capacity) on our flues, the arches also prevent the hot coals from accidentally entering the tubes and has also cut down on the number of hot cinders that come out of the stack and the amount of ash in the smokebox after each run. The 2-8-2 and 4-6-0 have two-piece arches and the 2-4-0, due to the smaller firebox door, has a three-piece arch. We have been very satisfied with how our engines are running with having an arch in the firebox.

Note: If I were to do this again, I would make the arch frame out of 316 stainless steel (so it will last better in the firebox) and use ⅛" thick material for the arch so it puts less stress on the frame.

Large Scale Atomizers

Figure 84. The Spirit of Traverse City is an oil-burning 15" gauge 4-4-2 "Atlantic" type steam locomotive that ran around Clinch Park Zoo from 1982 to 2011.

When Dad and I started working on the Spirit of Traverse City in the summer of 2004, she was a mediocre steamer. While it was possible to run her around the 0.4-mile loop while keeping the pressure between 140 and 150 psi (operating pressure is 150 psi), it was difficult to keep the water level up and not drop too much pressure. I was only able to do three trips around the loop before the water level would drop from just above half a glass to a quarter of a glass and it would take about five minutes to build the water and pressure back up. To help improve her steaming qualities we designed and installed a new blast nozzle, new oil pan and new atomizers.

For many years she had a single atomizer where the oil collected on a shelf above the steam jet. As the oil pooled, it would run (by gravity) towards the rear end of the shelf and drop over the edge where a jet of steam (.030" × 2.5") would spray the oil towards the rear of the firebox. While it did create steam, the fire was always a dark red and would smoke easily, no matter what blower or steam (for the atomizer) settings were. With this burner, it would be possible to have a spot fire, but as the burner was turned up, the fire would not fill the firebox evenly. The oil would not flow evenly across its shelf and the steam would not atomize all of the oil, so some oil would drip off the burner and burn on the bottom of the oil pan.

In June 2005, while the boiler was being retubed, we decided to build a new oil pan (to allow more air into the fire. It was also designed to be "easily removable" for cleaning and access to the burners) and try building new atomizers. I had seen Chet Peterson's design (Railroad Supply) for the burners that he used in his 1:8 scale 4-8-4 and based the larger burners off of his. As seen in my drawings (25 and 26), I did not scale his burners, but modified them to make them fit more easily in their home. Chet's burners were modeled from a round burner that I have seen used (with very good results) in 24" and 36" gauge engines.

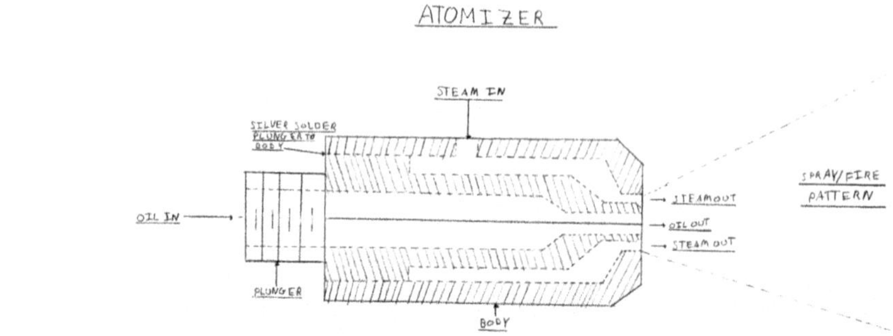

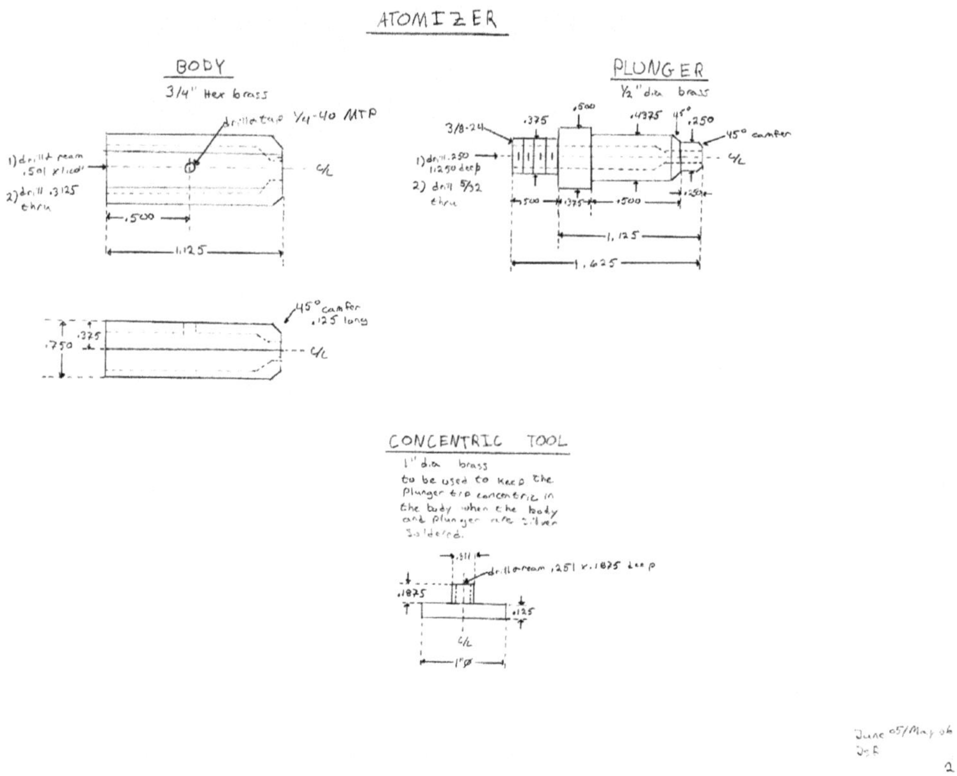

Top: Drawing 25. The oil enters the center tube in the back (left side of the drawing) and exits at the front of the center tube (right side of the drawing). The steam enters into the body around the center tube of oil and exits in a doughnut ring around the oil tube at the front of the atomizer. The steam creates a partial vacuum on the oil tube and helps draw oil from the oil tank as well as atomizing the oil. _Bottom:_ Drawing 26. The atomizer is made up of a body, the plunger/center oil tube and a "concentric tool" can help hold them in the proper alignment while they are being silver soldered together.

When building the burners it is important to make sure the plunger diameters are concentric to the oil hole and that the end of the plunger is line-in-line (flush) with the end of the body as well as concentric in the 0.312" diameter hole in the body. If the plunger and its hole are not concentric with the exit hole in the body, it is possible that the oil will not be evenly atomized and could drip oil. The concentric tool is used to keep the plunger concentric in the body while the body and plunger are being sliver soldered together.

The steam enters the body through a ¼-40 MTP hole. Drill and tap this hole (and debur) before the assembly is silver soldered. Any chips or debris could clog up the tiny opening (only a .032" wide circle) around the end of the plunger. The threaded end of the plunger is threaded ⅜-24 and this can either thread into a manifold for multiple burners, or into a fixture for a single burner. To seal the threads, I wrapped Teflon tape around the threads. Before I installed the burners on the manifold, I threaded a ⅜-24 jam nut onto the plunger threads. Once I had the burners rotated into the positions that I wanted, I ran the jam nut against the manifold to keep the burners from rotating.

With this being my first real experience with atomizers I didn't know how big the atomizer should be so I decided to use two of these atomizers. The burners are aimed 10 degrees apart, 5 degrees off the firebox centerline. I used ¼-40 MTP fittings leading back to ¼ compression and ⅛ NPT fittings for the steam manifold and used a 5/16" line to the steam valve in the cab. The manifold was drilled and taped for a ¼ NPT for the oil inlet and used ⅜" copper tube back to the oil valve. We used a ¼ NPT ball valve (located under the cab) with linkage up through the cab floor. The ball valve works quite well for controlling the oil flow.

Figure 85 shows the front section of the oil pan and atomizer. We welded a bracket

Figure 85. The atomizers are threaded into a bronze block. The oil line (bottom line) is threaded into the block which feeds the oil to the atomizers. I placed a ball valve in the block so I could clean out any debris that may collect there. The steam lines are above the oil line.

to the front of the oil pan and bent it to the "correct" angle. There is a large hose clamp that clamps down on the manifold, holding it to the bracket. The oil and steam lines are seen wrapping around the fireman's side of the oil pan, to where they will be connected with the rest of the piping. I also added a ¼ NPT ball valve onto the manifold as a drain valve in case a lot of dirt or debris were to get into the oil line. This allows us to drain the line and not blow it through the atomizers.

After several years of use, we have had to make several adjustments to the burner and oil pan. The burner has been adjusted to a higher angle as the lower angle was causing the spray to hit the bottom of the pan. The shield over the burners has been removed and the burners are now aimed towards the firebox door.

What is not seen in figure 86 is the piece of firebrick that is just inside the door, standing just below the door along the door sheet. This brick acts as a flash wall when the fire is sucked out (or blown out by water in the steam or oil lines) and for the most part, is able to ignite the fire after it has gone out.

Figure 88 shows the flame as seen from the front fireman's side of the box. The ends of the burners are visible near where the flame starts. These burners produce a hot, yellow-burning flame that fills the firebox.

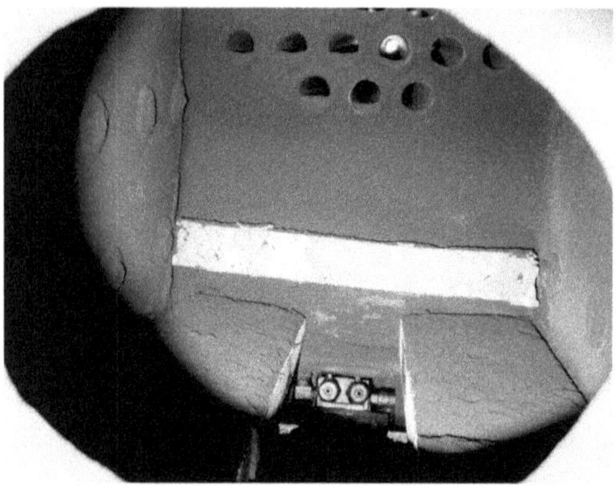

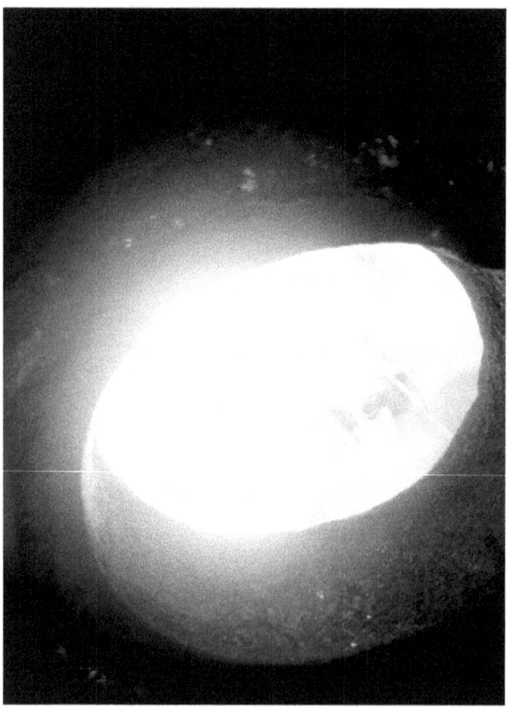

Top left: **Figure 86. This picture was taken the morning after the tubes had been brushed out; the carbon dust on the arch bricks is from brushing out the tubes. The atomizers are angled up towards the door.** ***Bottom left:*** **Figure 87. The atomizers produce a bright yellow flame fire that produces a lot of heat. The fire in this picture is of a medium to low fire, the setting that would be used when drifting or sitting at the station. When the fire is turned up for climbing a hill, the entire firebox is filled with bright yellow flames.** ***Below:*** **Figure 88. The atomizer is mounted below the throat sheet. While the atomizer is not seen, the fire is seen.**

The steam passing over the end of the plunger creates a partial vacuum which helps to draw the oil through the burners. With the new burners, oil pan and blast nozzle, I was able to do five trips around the 0.4-mile loop without stopping and I was able to keep the pressure between 140–150 psi and the water between one-third and one-half a glass. There was a 300' or so long climb up a 2.5 percent grade and a curve around 100–110' radius. While pulling three cars loaded with passengers (2,000–2,500 pounds each) using the old burner, the boiler would usually drop 10 psi. With the new burners (four notches out of center, blower off, burners set) she could hold a steady 145–150 psi and when the tubes were clean, she would gain a few psi.

Two burners worked well in this 13" × 18" firebox. If there would be any really long grades, a third burner might be needed (or the atomizer size increased) as with these two, it was only on a rare occasion that water could be added while working up the 2+ percent grades. This size burner could most likely work well in the large 1.5" scale engines or in the 2.5" narrow gauge engines. Because they produce a good, hot fire, care must be taken so the boiler is not fired up too quickly. A good rule of thumb is that it should take one hour per foot diameter of the boiler to fire up from a cold start to full operating pressure. With a solid fuel fire, this is pretty easy to do as it takes a while for the solid fuel to start producing a hot fire. With oil, it is really easy to get a hot fire, so a spot fire is what is needed when starting up. The Spirit of Traverse City had a 16–17" diameter boiler. We would aim to have the pressure gauge off the peg about 35 minutes after lighting the fire and have full pressure about an hour and 20 minutes after lighting the fire. This helped minimize the stress on the firebox sheets, staybolts and tubes.

The Spirit of Traverse City used these burners from 2005 to 2009. In 2009 new burners were made, as the older ones were getting dirty and needed to be cleaned. She used the replacements until her last trip in 2011. After installing these type of burners, the main reason for not being able to keep up steam was because the flues were dirty. These burners worked quite well and help up well with around 100+ days of constant use a year.

Drafting

While I was a student at Michigan Technological University, one of my classes was called mathematical modeling. One of the projects I chose to do was to model the drafting (blast nozzle size, stack size, etc.) of a steam locomotive. In order to help simplify things I decided to look at a single port exhaust nozzle. Not only was I able to use the information I gleaned from this project for the engines that my dad and I had built but also I was able to use it to improve the drafting on the 1:4 scale 4-4-2 that used to run in Traverse City.

Single Port Blast/Exhaust Nozzle

The earliest forms of the blast/exhaust nozzle was the single port exhaust nozzle which basically was just a pipe carrying the exhaust from the cylinders and directing it up the stack. As the engines became larger, so did the stacks and the style of the nozzle had to be changed to make sure the exhaust steam would fill up the stack. For most live steam engines, the single port nozzle will work very well. What I will be describing is how I designed a nozzle where you can change out the nozzle easily to try different sizes and find the one that works best for your use.

Also, this nozzle will work for geared engines (such as Shays, Heislers, etc.) and any rod engine that has only one blast nozzle and one stack where the exhaust blasts are separate from each other. With an articulated with one stack and all high-pressure cylinders, it is possible to have the wheels slip to have more than one exhaust blast at a time. With compound engines having a large low-pressure cylinder and lower steam pressure in that cylinder, I don't think these calculations will give the same results as with a simple engine with separate blasts.

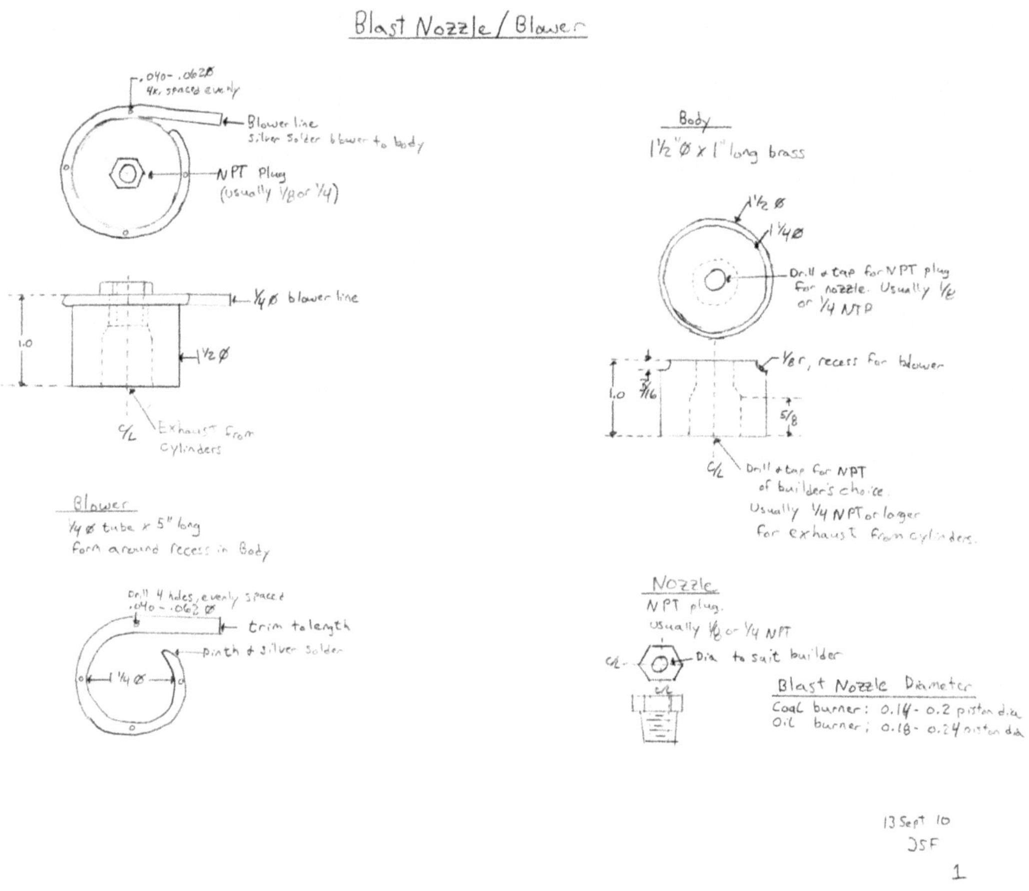

Drawing 27. The blast nozzle I designed uses a pipe plug with a hole drilled in it for ease of changing blast nozzle sizes and has a ¼" OD copper tube soldered to the blast nozzle for the blower.

I designed the blast nozzle in drawing 27 to also include a blower line made of ¼" copper tubing. The builder can decide whether to use hard or soft copper and how to plumb it. The body of the nozzle is made out of 1.500" diameter brass (bronze will also work) and the nozzle is a drilled-out pipe plug.

The smaller the nozzle is, the more velocity the exhaust steam has going up the stack and the stronger the draft is on the fire. The drawback is that the back pressure (the pressure of the exhaust steam in the exhaust passageways and on the exhaust side

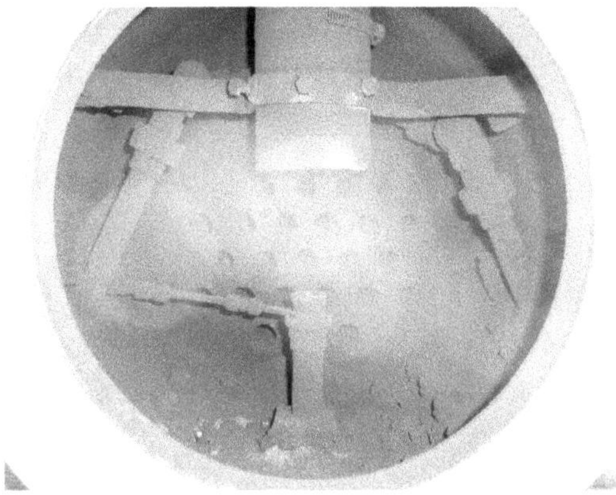

***Above, left:* Figure 89. This is the front view of the blast nozzle in the 2-8-2 that my dad and I built. The ¼" line with the 90-degree compression elbow is the blower line. The blast nozzle is sitting almost on the floor of the smokebox to give the most distance between it and the petticoat. *Above, right:* Figure 90. This is the top view of the blast nozzle showing the blower ring, the four blower holes, the nozzle, and in front of the nozzle is the throttle (a ½ NPT ball valve) and the steam pipes to the cylinders. *Right, bottom:* Figure 91. This is the new blast nozzle and blower I designed and made for the Spirit of Traverse City. What I found in doing research about the single port nozzles is that for a coal burner the nozzle diameter is 0.14–0.20 the diameter of the cylinder. For oil burners the nozzle diameter is 0.18–0.24 the diameter of the cylinder.**

of the piston) is higher. The larger the nozzle is, the less velocity the exhaust steam has and the less draft there is on the fire. Also, back pressure is less.

For a piston that is 2.000" in diameter, in a coal burner the exhaust nozzle will be 0.280"–0.400" and in an oil burner the exhaust nozzle will be 0.360"–0.480". From my experience most of the live steam engines are not working that hard and are not going that fast so the exhaust nozzle tends to be on the smaller end of the range.

The sketches in drawing 27 show how I made the nozzle and the different pipe sizes are left up for the builder to decide what will work best on their model and different nozzles (label them with their size) can be made up and changed easily throughout the running day to see how the engine responds to the different sizes.

Note: The ranges in blast nozzle sizes that I found for coal burners (14 percent to 18 percent of the piston diameter) and oil burners (18 percent to 22 percent of the piston diameter) were from texts from the late 1800s to early 1900s. By the 1920s railroads and builders had settled in on a round blast nozzle being 22 percent the diameter of the piston. The engines in those days were working up to their maximum tonnage and due to the large volume of steam being used and passing through the blast nozzle created a good draft. Many steam locomotives today are operating well under their maximum tonnage and with not as much steam being used, the draft is less. Some engines have had their blast nozzles made smaller to create a better draft when pulling less tonnage at slower speeds. Many live steam engines are not worked to their full capacity so a blast nozzle that is 22 percent of the piston diameter may not create enough draft.

Smokebox Set-up

In order for a good draft to be formed, the exhaust nozzle and the stack must be in line with each other so the exhaust blast will be centered in the stack. The nozzle must be the correct size so the exhaust blast will fill up the stack to create the partial vacuum in the smokebox.

The smokebox must be airtight. Seal off the open space around the blast pipe with furnace cement and seal off any other openings around the steam delivery pipes, where the smokebox joins the boiler and the smokebox door and stack. If any air is allowed to enter the smokebox, it will weaken the partial vacuum and will lessen the draft on the fire. Also, if there are cinders in the smokebox, the air may allow the cinders to burn. Additionally, there should be no steam leaks from any of the pipes (including the exhaust pipe) as the steam will partially fill the vacuum in the smokebox and lessen the draft.

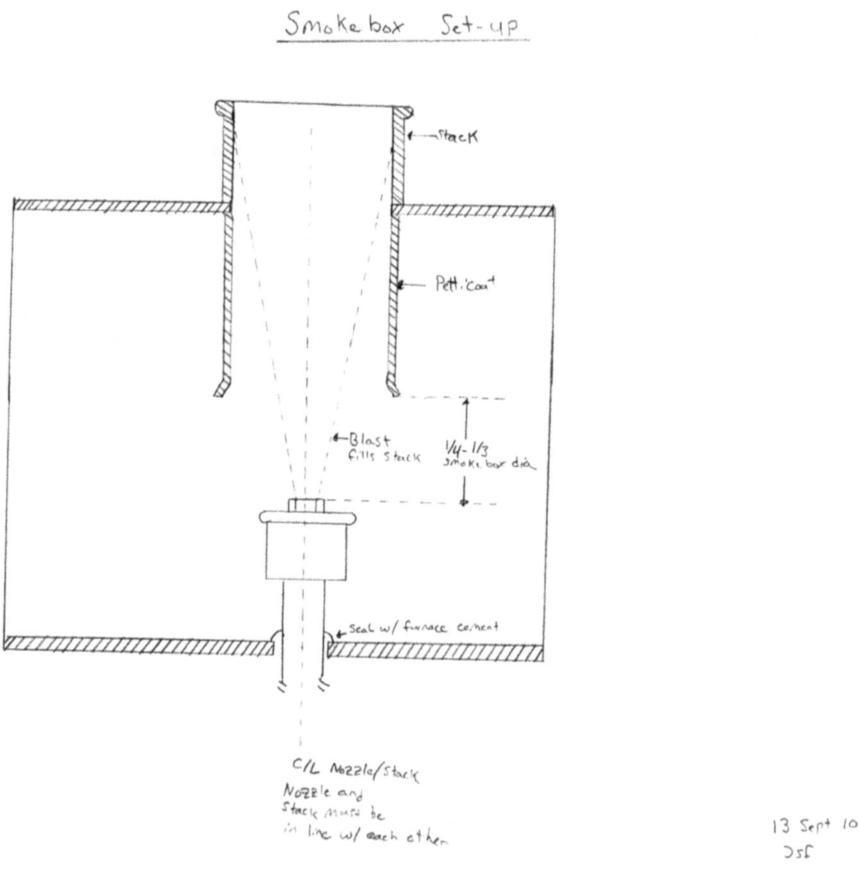

Drawing 28. This shows the basic placement of the blast nozzle and blower in the smokebox.

On live steam engines, I have found the best results placing the exhaust nozzle as close to the floor in the smokebox as possible and extending the petticoat pipe down rather than moving the nozzle up. The best distance between the petticoat pipe (we have had good results with a flared petticoat and a straight walled petticoat) and the nozzle has been between one-quarter and one-third the diameter of the smokebox; on most 1:8 scale engines, that is around 2"–3".

Some other modifications can be done easily at this point. If a coal burner tends to throw hot sparks from the stack, you can wrap screen around the petticoat pipe (secure it with a hose clamp) and extend it to the floor of the smokebox. We did this on our 10-wheeler and found that 3⁄16" screen will allow plenty of draft while stopping a lot of hot cinders from flying out of the stack.

Figure 92. The spark arrestor in the 4-6-0 is made out of 3⁄16" wire screen.

Also, if the engine will be converted to propane, it will need to have less draft, as unlike coal, less draft equals more heat. What we have done on our 10-wheeler is installed an extension to the petticoat pipe made of sheet metal and secured it with a hose clamp. We would shorten the distance between the blast nozzle and the petticoat to ½" or less and it worked very well.

The Firebox

The other aspect of having an engine draft and fire well is in the firebox. Most of my experience is with coal burners and what we have found is that the fires need more air coming into the fire than what you would get by scaling down the grates of a full-size engine. You generally need at least a 1:1 ratio of solid grate area to air area (bar versus the space between bars). That means if your grate bars are ⅛" thick, the space between the bars must be at least ⅛". We have found that on our 2-4-0 we need a ratio of 1:1.67 to have the engine fire well. That means our grate bars are .187" thick and the space between bars is .312". On our larger engines, .250" grate bars and .312" space between them works well.

The only experience I have had with an oil burner is with the 1:4 scale 4-4-2, the Spirit of Traverse City, and we had to increase the size of the oil pan and increase the air flow into the fire. My understanding of oil is you want to have the largest firebox volume that is reasonably possible and you need to have a smaller air opening in the front by the burner with larger air openings in the rear of the pan (in a large firebox along the sides, too) and have air flowing through a bottom vent in the firebox door. If I remember correctly, the total area of the openings should be 75 percent to 100 percent of the area of the flues. If there is too much air flow at the front by the burners, the air will take the shortest way to the flues and instead of following the path of the fire it will jump up from by the burners to the lower flues (the draft on the front of the firebox is the strongest) and thermal shock the flues which can cause the flues to loosen in the tube sheet and leak.

Personal Experience with Axle Pumps and Injectors

In live steam there appear to be several ways that are used the most to put water into the boiler. Axle pumps, injectors and steam-driven water pumps are the ones I have seen most often used. I have heard people say that the injector is more trouble than it is worth and using an axle pump or a steam-driven water pump is much more reliable. On the other hand, some people say that their injector works great and they only use the injector and just leave the axle bypass open.

When Dad bought the Shay (seen in figure 1) in late 1989, she had an axle pump on the lead axle of the lead truck and a hand pump in the tank over the rear truck. When we ran the Shay, the axle pump did not seem to keep up with the demand for water so Dad made a new larger plunger and a new body with a larger bore. For the most part, the axle pump did work properly (aside from dirt getting in the check valves) but when the pump pumped the water, there was enough force to cause the truck to shift/twist in rhythm with the pump.

When we built the 0-4-0 (Railroad Supply), we installed an axle pump that was driven off of the main axle. We had the axle pump pull the water through the hand pump (which ended up being helpful later on) and installed an injector from Coles on the fireman's side. The 0-4-0 ran at the last meet at Bill Fitt's track in June 1996 and the Coles injector did not want to work above 100 psi and it worked only marginally at lower than 100 psi. We had installed two horizontal ¼-40 MTP checks in the delivery line and we now believe that since these checks are not very free flowing, that there was too much resistance in the line to allow the injector to work freely. We also found out that one of the stainless steel balls used in the suction check in the pump was too small and would get stuck and not allow water to enter the pump. Luckily, since we had the hand pump plumbed through the axle pump, we would use the hand pump to free the sticking check valve.

Figure 93. The 0-4-0 is running, but not complete, and had the injector removed when this picture was taken. Howard Alton Frost, my grandfather, is at the throttle. He loved trains and steam locomotives. This was the first steam locomotive he had ever run. This was taken at the White Creek Railroad in July 1996 (courtesy Donald S. Frost).

By July 1996 the injector had been removed, the faulty check valve in the axle pump had been replaced and the hand pump had been plumbed separately from the axle pump. While the axle pump did keep up with the water demand, it was very difficult to keep the pressure up. Part of the problem was our lack of firing skills (the Shay burned oil in a pot burner; the 0-4-0 was our first coal burner) but we also did not have a fine enough bypass valve to finely adjust the bypass to allow just enough water to enter the boiler and the rest to head back to the tender tank. We also noticed that when the pump was pumping water into the boiler, the engine would slow down with each pump. To help our pressure problem, I tried installing a feedwater heater coil in the smokebox. I used about 12' of ¼" copper tubing and wrapped it in a coil in the smokebox. I cut the delivery line as it came out of the pump and connected it to the coil and then connected the coil to the rest of the delivery line. It did help to heat the water but not enough to overcome the pressure problem. During this time we tried several other injectors (we were told that they were English injectors, but we had no idea what kind or type they were) but we could not get them to work well, especially above 100 psi. The engine operates at 125 psi and if we came back from more than a half-mile run with 70–80 psi, we thought we were doing great.

Around 1997 or '98 Dad bought a Chicago-style injector from SuperScale for the 2-8-2 that we were building. We decided to try it out on the 0-4-0. When we first piped it up, we used it as a lifting injector, but it has since been moved below the water line in the tender due to the injector sometimes getting too hot to pick up water when we were using it as a lifting injector. When we tried it, it worked great. When running, we left the axle pump bypass into the tender and just used the injector. The engine did not have near the pressure problems (we later installed a stainless steel arch) and were able to keep the pressure between 120 and 130 psi. By late 1998 we had disabled the axle pump.

We left the pump body in place "just in case" and because leaving it was easier than removing it. We later added a second injector (SuperScale's economy injector) to the engineer's side.

Dad and I started building the 2-8-2 (Railroad Supply) in late 1996 and she first ran in May 1999. When we came to the axle pump, we decided to make and install the bracket as it also acts as a frame spreader.

We installed the eccentrics on the original main axle as a "just in case" but never used them. (When Dad made new driving wheel sets, he did not install the eccentrics for the pump.) We had decided that we would use injectors on the 2-8-2 and then later on the 4-6-0 when we built her.

We started the 4-6-0 (Allen Models) in September 2000 and she was finished in May 2001. We did not do anything for an axle pump on this engine but decided to use two economy injectors from SuperScale. All of our engines have a hand pump (one from Railroad Supply, one from LocoParts and one homemade one) to use for a hydro test, but that is all we have to use it. Aside from an occasional cleaning, the injectors have been working great.

Our tenders were fabricated by one of Dad's friends and are made out of galvanized sheet steel. The tank for the 2-8-2 was made in 1999 and the ones for the 2-4-0 and 4-6-0 were made in 2001.

We have been very pleased with the performance of SuperScale's Chicago and Economy injectors. They work very well throughout the operating pressures (and fire up pressures) and are easy to use and maintain. I like to alternate which injector I use to even

Left: Figure 94. When we disabled the axle pump, we removed the plumbing to the pump but left the pump in place on its bracket. The bracket also acts as a frame spreader. *Right:* Figure 95. The axle pump bracket was installed on the 2-8-2 when we built her as it also acts as a frame spreader.

Figure 96. The 2-8-2 has two Chicago non-lifting injectors. The engineer's side one is pictured. These injectors use 5/16" tube for the steam and delivery lines and 3/8" tube for the water line.

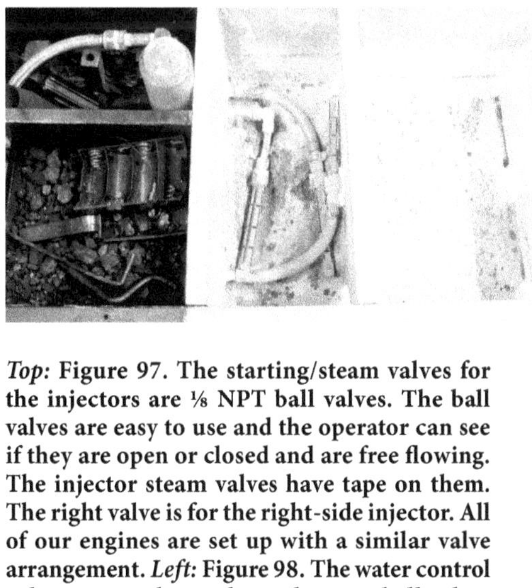

Top: Figure 97. The starting/steam valves for the injectors are ⅛ NPT ball valves. The ball valves are easy to use and the operator can see if they are open or closed and are free flowing. The injector steam valves have tape on them. The right valve is for the right-side injector. All of our engines are set up with a similar valve arrangement. *Left:* Figure 98. The water control valves are on the tender and we use ball valves with a modified handle to control the water flow. To the left is the arrangement for the right-side injector on the 2-4-0. *Above:* Figure 99. The filters we are using are inline garden hose filters. They have a fine mesh screen and a large surface area to stop particles but not impede the flow of the water. These are the original filters that we installed. ⅜" copper tube is used for the water delivery lines for the injectors and handpump. This is in the 2-4-0's tender. We did not treat the inside of the tank and in this photo (taken October 2014) there are some rust spots that were cleaned up and treated over the winter of 2014–15.

out their wear and to make sure that they both are working. We have not had to use the hand pumps as an emergency backup as even if one injector stops working (dirt or scale in the injector is the typical reason) or if the delivery check on one side fails, the other injector has been working like it should.

Tires

Note: The information in this section can be adapted to any steam, diesel, electric or gas locomotive of any gauge.

The practice for steam engines is to put a steel tire over the center, or the hub of the wheel. The hubs are generally cast iron for the earlier engines and then cast steel as technology progressed. As the engine would run, the rail would wear down the tire and if the tread was hollowed out too much or the flange was worn too thin, the wheels would be dropped and the tires would be turned and a new profile would be machined. Once the tire was too thin, the tire would be heated up and slid off the hub, a new tire made, heated and slid back over the hub. When the tire cooled, it would have a tight grip on the hub. Tires can be also applied to car wheels or diesel wheels so the axles and centers don't have to be replaced.

***Above:* Figure 100. This is one of the wheel sets from Strasburg Rail Road #7312. Her drivers were dropped out and the wheels were turned on our 90" Niles wheel lathe. *Right:* Figure 101. This picture shows the new profile that was machined. Between the two black lines is where the diameter is measured. The tread has a radius on the inside and outside, there is a full fillet between the flange and the tread, the tread is tapered and there is radius on the outside of the tire to the outside face.**

Figure 102. Here #89 is being fired up for a day's run and the tires on the fireman's side can all be seen. The riveted ring is a retaining ring to keep the tires from coming off the wheel center if the shrink fit failed. The tire is shrunk, not riveted on.

During the summer of 2001 Dad and I noticed that something was wearing against the inside edge of the flanges and that they were getting thin. (As we later discovered, we had machined the driving boxes to the wrong dimension. Instead of machining the width of the boxes for 7.5" gauge, we had machined the width for 7.25" gauge which gave the wheels too much lateral movement, allowing them to shift against different bolts and bolt heads that were holding things to the frame.) So we decided to drop out the wheel sets, turn the drivers down and add steel tires to the hubs. It worked so well on the Mikado that we decided to add steel tires to our other engines, the 2-4-0 and 4-6-0. All of the engines had cast iron wheels and one of the biggest differences we noticed was with the 2-4-0. With cast iron wheels (cast iron is great for bearing material, not necessarily for traction) the engine was quite slippery and it was hard to pull a train of six cars up a decent grade. With the steel tires, she did not slip as much and I have been able to pull nine cars up a 2 percent grade easily, without slipping or losing pressure and water.

Figure 103. The tires can be easily seen on the 2-4-0. On our Mikado they are painted and blend in more with the hubs. The tires on the 4-6-0 are easier to see as they have been painted white.

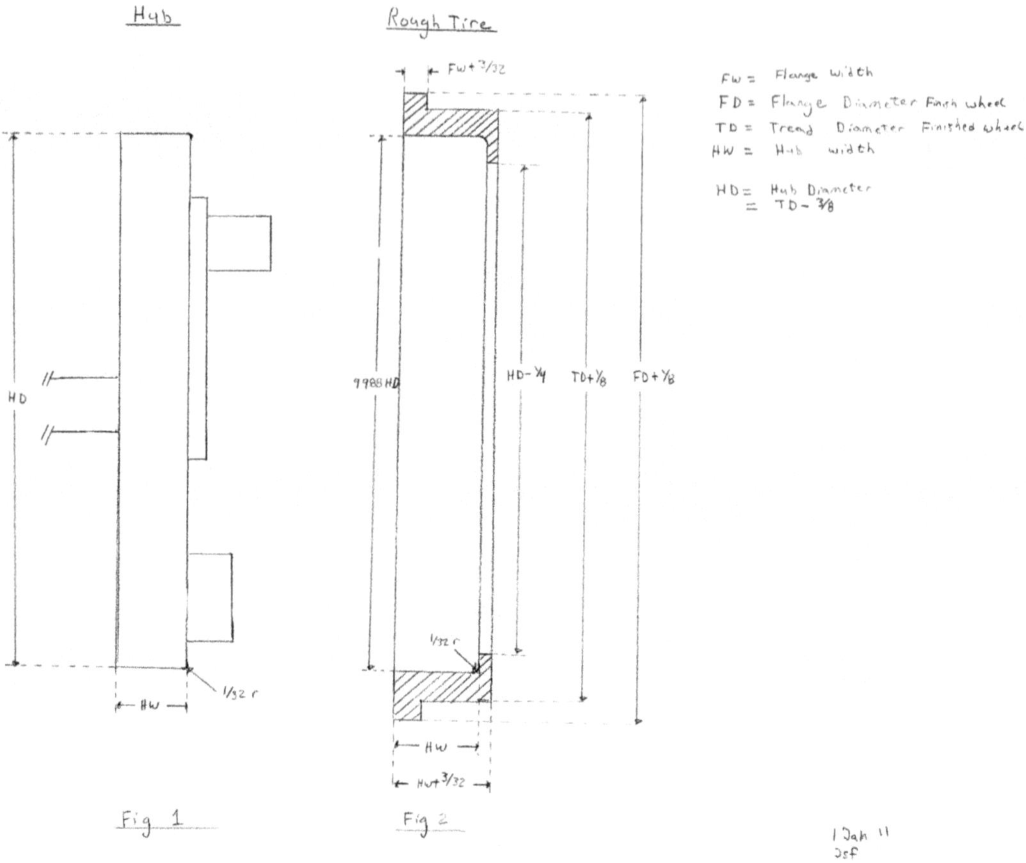

Drawing 29. Figs. 1 and 2 show a general formula for the dimensions and for machining the hubs and tires. This is to allow tires to be made for different size wheels for 1:8 scale as well as for other scales. The generic dimensions are as follows: FW = flange width, machined; FD = flange diameter, finished wheel; TD = tread diameter, finished wheel; HD = hub diameter (which for 1:8 scale is TD−.375).

For 1:8 scale, the wall tire thickness should be no less than 3⁄16", meaning that the hub diameter is 3⁄8" (at the minimum) smaller in diameter than the finished tread diameter. (If the tire thickness is increased, the tire life can be increased before the tire becomes too thin. For 1:8 scale, a ½" thick tire should be able to be worn and returned before it becomes too thin. The thickness of the tire also depends upon how thick the rim of the wheel is. The rim should also have a bit of meat left on it [I would not go thinner than ¼"–5⁄16" for cast iron wheels] so it doesn't break and can support the tire.) For larger scale engines, the wall thickness should be increased.

When the hubs are turned down, it would also be a good idea to face the outside edge of the wheel to make sure it is running true to the OD of the hub and to clean up any paint, oil or grease. Also, a 1⁄32" radius should be machined at the corner of the OD to the outside face because when the tire is bored out, there will be a small radius (typically 1⁄32") left by the tool where the bore stops at the lip. The lip on the tire is to give the tire a stopping point so it won't slip off the wheel as it is cooling and contracting onto the hub.

As a general rule, for every inch in diameter the hub is, the bore of the tire should be 0.00125" smaller than OD of the hub; or the bore should be 0.999–0.99875 the diameter of the hub.

Extra material is left on the tread and flange services. This is to allow for the final turning of the tread once the tire is on the hub. There is a chance that the tire might not sit square or that the tire may not run true to the stresses of being heated and squeezing on the hub. However, if extra care is taken to make sure that the tire is sitting square on the hub and the gauge is correct, the tread can be finished before the tire is installed on the hub.

To install a tire on full-size engines, the tires are heated by having a pipe wrapped around the wheel. On the wheel side of the pipe are small holes to allow gas (sometimes propane or a mixture of propane and compressed air) to burn and heat up the tires to about 450°F. To heat up the large tires, it generally does not take more than 13–15 minutes. For live steam models, either a propane grill or an oxy-acetylene torch with a rosebud tip will work. For using the grill, place the tires on the grates (multiple tires can be heated

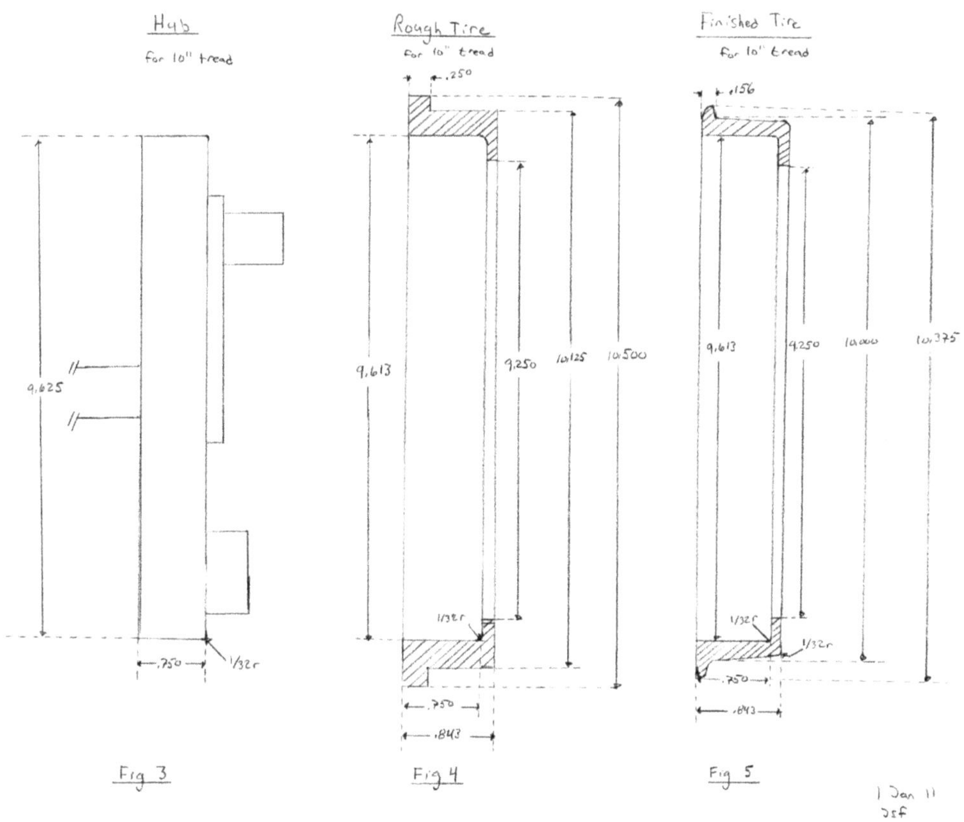

Drawing 30. Figs. 3, 4 and 5 show the dimensions that could be used with a 10" diameter driver typically found with a 1:8 scale 4-4-2, 4-6-2, 4-6-4 or 4-8-4. The dimensions are not set in stone, but care must be taken to make sure that there is the proper amount of interference between the hub OD and the tire ID. If there is too little interference, the tire can come loose easily. If there is too much, either the tire or hub may break.

and installed easily) and turn the grill on high. Typically, after about 15–20 minutes, they are hot enough that they have expanded to be able to drop over the hub. If heating with a rosebud torch, heat the tire evenly and don't over heat any one area. *The tires should* not *turn blue as that is way too hot. The tires should be heated to a dull straw color, which is around 400–450°F.* When putting the tires on the wheels, they should be set squarely over the hub as they can easily bind if they are slightly cocked. The tires will also cool quickly so if the tire is too snug to begin with, heat it up some more. Also, have a hammer nearby so as to give them a quick tap to help the tires seat. Once the tires have cooled, they will have a good grip on the hubs and can be turned to their final dimensions

Notes on fitting tires:

1. The lip does not need to be as tall as shown in the sketches of drawing 30. A ⅛" tall lip will work well.
2. I have machined or sweated on and off over 100 tires ranging from 4" in diameter to 80" in diameter. If the shrink fit/interference fit is good, the tires won't come loose.
3. The tires can be finished before or after they are sweated onto the center. If they are fully machined before they are sweated on, care must be taken to make sure they properly seat on the wheel and are square to the wheel.
4. The profile of the tires should follow IBLS standards (there is only a minimum width, no maximum width, maximum flange height or maximum flange width) which can be found here: http://ibls.org/mediawiki/index.php?title=IBLS_Wheel_Standard.

Automatic Cylinder Cocks

Around 2000 I started messing around with automatic cylinder cocks that can be controlled from the cab. I had seen several engines that had commercially available steam powered automatic cylinder cocks, but we didn't feel like spending the money to buy those so I decided to make some. The cylinder cocks are opened and closed by valves in the cab. To close the cylinder cocks, the drain valve is closed and the valve on the turret is opened to pressurize the line to the cylinder cocks. If for whatever reason the pressure in the cylinder exceeds the pressure to keep them closed, they will open and release the pressure. To open the valves, the valve on the turret is closed and the drain valve is opened. Over the years the dimensions have changed and have been modified for the different engines that Dad and I built, and I made larger ones that we used on the 15" gauge 4-4-2, the Spirit of Traverse City, which ran around the zoo in Traverse City, Michigan.

I made the body of the cylinder cock out of ½" brass hex as that is what we had. In order to connect the body to the cylinder block, I drilled and reamed a ¼" hole to silver solder a hard ¼" tube and then threaded the tube for a ¼-40 MTP thread. The holes in the bottom of the cylinder blocks are ⅛ NPT so I took ⅛ NPT plugs and drilled and tapped them for ¼-40 MTP. I designed it this way as the ¼" tube will shear off in a derailment instead of having the cylinder cock pull out or damage the cylinder casting.

The plunger is made out of ¼" brass. The 0.075" diameter end keeps the cylinder cock centered and aimed for the exhaust hole. To seal the plunger against the exhaust hole, I held the body in the vise and tapped the plunger against the exhaust hole. The

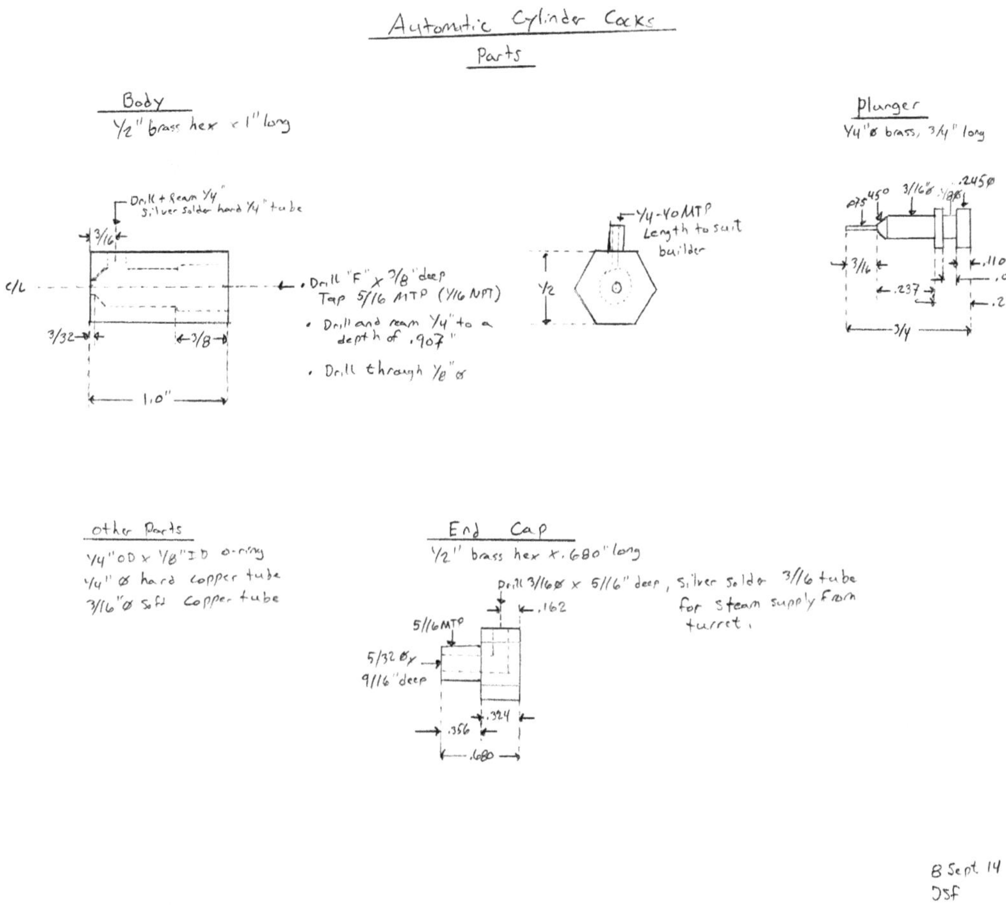

Drawing 31. This is the generic drawing I used for making the cylinder cocks.

Left: **Figure 104. View of the completed cylinder cock showing the outside face where the steam will vent from the cylinder.** ***Right:*** **Figure 105. Side view of the cylinder cock showing how the end cap connects to the cylinder cock body.**

Figure 106. The parts of the cylinder cock (body, plunger, O-ring, end cap) are laid out.

end cap end is sealed by a ¼" OD × ⅛" ID O-ring. We have just used common O-rings that are available at the hardware store. In 14 years, we have only had to replace 4–6 O-rings between all three engines.

The end cap does not have to be made in this way if there is room for commercially available pipe fittings. When making changes to how the end cap is applied, the threads should not extend to where the O-ring will seal against the walls as it will leak and chew up the O-ring in short order. The end cap should also thread in far enough to keep the plunger from dislodging from the ⅛" exhaust hole. If the 0.075" end is allowed to fall into the body, it will take a bit of working to fish it back out.

To close the cylinder cocks, the exhaust valve is shut and the steam supply valve is opened. This will push the plunger against the exhaust hole, which will prevent steam from escaping from the cylinder. The O-ring will prevent the steam from the turret from entering into the cylinder.

***Left:* Figure 107. This is the set-up on the 2-4-0. There is minimal clearance under the cylinder block so the cylinder cocks and piping are tucked in under the casting. The plate going under the cylinder cocks is a skid plate used to take most of the force in case of a derailment to help save the cylinder cocks. *Right:* Figure 108. The piping arrangement had to be slightly different on the 4-6-0 due to the lead truck.**

To open the cylinder cocks, close the steam supply valve and open the exhaust valve. When the pressure is drained from the line that runs between the turret and cylinder cock, the steam from the cylinder will push the plunger back and exhaust through the exhaust hole. If there is more pressure in the cylinder (like if the cylinder is full of condensation) the cylinder cocks will act as safety valves and open automatically as the pressure from the cylinder overcomes the pressure, keeping the cylinder cock closed so the pressure will be released and not damage the cylinder.

Figure 109. Cylinder cock control valves in the 2-4-0's cab: (1) steam supply valve from the turret; (2) exhaust valve (I am washing the boiler in this picture). On the 2-4-0 these valves are ¼ MTP valves. The other two engines use ball valves.

Replacing a Crank Pin ... and Then Some

While I was doing an inspection of our 4-6-0 and making a few minor adjustments, I noticed that the main right-side crank pin was loose in the wheel center. Normally the crank pin would be pressed into the wheel center with a large enough press fit so it would not come loose. With the crank pin being loose, either we did not get a good press fit or the wheel center was cracked, allowing the hole to open up. Since I don't have my own machine shop (I do have access to a machine shop which I used when making the new pin) or any way to get the wheel set out from the locomotive, I had to figure out how to get the pin out while the wheel was still under the locomotive.

When we were building the 4-6-0, instead of just relying on a press fit (or our attempt to get a good press fit) to hold the crank pins in the wheel centers and wheels on the axles, we drilled and tapped a ¼-28 threaded hole and ran a set screw into the hole. The hole was drilled so half was in the wheel center and half was in the crank pin or axle. This way if a pin or axle came loose, the pin and/or wheel would not just fall out. The crank pin was loose, but this set screw kept the pin from rotating and it also meant I had to figure out how to get that set screw out.

With the Railroad Supply 0-4-0 (modified into a 2-4-0) and 2-8-2 we can gain access to the crank pins if we have the crank pin at the 6 o'clock position and remove the pedestal binder. We could then use a rod and drive out the offending crank pin. However, the Allen Models 4-6-0 uses a built-up frame where the bottom frame rail runs the entire length from the cylinders to the rear frame casting. This, coupled with how the spring rigging is attached, meant that there was no easy access to the crank pin when the pin was in the 6 o'clock position. I did notice that there was ample clearance between the top of the journal/driving box and the top frame rail, but I would need to disassemble part of the spring rigging to do this.

Top: Figure 110. I removed the pilot (cowcatcher) and set the pilot beam up on blocking. *Bottom:* Figure 111. I made a holder to support the rear section of the frame so the wheels would be off the rail/stand.

I removed the pilot and used wood blocking under the pilot (figure 110) and under the rear of the locomotive frame (figure 111) to allow the driving wheels to be off the stand. Since the engine is not too heavy, I was able to pick up the front quite easily and slide the blocking underneath. The rear was a little more interesting as it was hard to find an easy way to hold onto the engine to lift it and then push the blocking under the rear of the frame. Since the rear of the frame is tapered, once I got the wood under the frame, I was able to use a hammer and drive the wood under the frame which would then pick the frame up.

With the wheels off the rail and the journal/driving boxes bottomed out in the frame, there was still tension on the spring rigging so I could not just disassemble and get part of the spring saddle hangers out of the way. In order to get some slack in the spring rigging, I took a vise grip and clamped on the front of the spring that was behind the main driver. This put tension on the rear driver and loosened up the rigging over the main to get the pin out and lift the hanger clear of the top of the box.

***Left:* Figure 112. I used a pair of vise grips to clamp the front end of the spring behind the R2 driver to the bottom frame rail to loosen the equalizing over the R2 wheel. *Above:* Figure 113. With the spring rigging loosened up by clamping the spring behind R2 I was able to remove the pin in the front of the R2 spring saddle hanger.**

To remove the set screw from the back of the crank pin, I placed the right side crank pins at the 12 o'clock position to gain access to the set screw above the driving box.

To drive the crank pin out, I again gained access above the journal box by using a ⅜-16 long piece of all thread against the pin and hammering on the all thread from the left hand side of the engine.

Figure 115 shows the right main driver with the crank pin removed. I cleaned up the hole and did not see any cracks (good) which meant that the press fit was non-existent (bad).

A small problem showed up during this process that was then also fixed. When we were building the 4-6-0 we tried to harden different pins ourselves. We were very successful in hardening the pins, but not so much with the tempering. We didn't learn this

Left: Figure 114. In order to remove the set screw keeping the crank pin from falling out, I needed to rotate the wheels so the crank pin was above the drive box so I could get an Allen wrench into the set screw. Right: Figure 115. I have removed the main crank pin from the wheel.

aspect until some of the hardened pins sheared off. Since we really didn't get a good temper, the pins were hard but brittle.

Figure 116. In addition to needing to make a new crank pin, I need to make a new eccentric pin as the old one was broken.

It was about this time that another larger problem showed up. The right main and right rear wheels were loose on the axles. The set screws were keeping them from sliding off the axles and the keys were doing the main driving and keeping the wheels quartered. I was hoping that this meant that the press fits between the wheels and axles were bad and not that the tram (spacing of the axles and the side rods) was off, thus loosening up the wheels. I decided to remove the wheels, clean them up and inspect them to make sure the castings and axles appeared to be in good shape. In order to remove the wheels, I removed the set screws, disconnected part of the brake linkage and then had to hold the brake shoes clear of the flanges and was able to pry the wheels off.

Top: Figure 117. Due to both the R2 and R3 wheels being loose on the axle, I removed the wheels and cleaned the axles and the bores. *Bottom:* Figure 118. The right main (R2) and right rear (R3) wheels are lying next to the right side and main rods.

I machined a new eccentric crank pin and a new main crank pin and pressed them into their respective homes. I made the new pins out of 4140 and I turned the press and journal diameters about .004–.006 larger than their final size. I then used 80 and 120 grit Emory cloth to polish out the machine marks and to get a good, smooth final size. Just leaving it as a machined surface, the tool will leave marks that will have high and low spots which will cause it to wear more until everything is smooth and for a press fit, it will grab more on the peaks and not as much (if at all) in the valleys.

A general rule of thumb for a press fit is about 0.00125–0.0015 of press per 1" of diameter. Because the crank pin was loose, the bore in the wheel was about 0.0015" out of round and tapered by about 0.001" and since I didn't have a good way to rebore the hole and keep it in the same location, I went with an average press fit of 0.0015"; with some of these smaller parts, there is enough meat around the hole to allow a larger press fit. If I left the machine marks in the crank pin, the peaks could be pushed over into the valleys and I would not get a good fit. I also turned the first 1⁄16" of the press fit diameter 0.001" smaller than the hole to allow the pin to start straight into the wheel.

***Left:* Figure 119. The new main crank pin is pressed into the wheel and the new eccentric crank pin has not yet been pressed into the eccentric crank and is lying next to the wheel. *Right:* Figure 120. This is the back side of the new R2 crank pin. The crank pin is pressed in and the set screw will not be used.**

There was no sign of trauma (metal shavings, cracking) on the back side of the wheel and I decided not to use a set screw but instead used Loctite 603 as a backup to the press fit; I don't believe I needed to, but there would be no harm in also using Loctite.

While the wheels were off the axles, I cleaned them up and did not see any cracks or any trauma in the bores or on the axles. The bores measured 1.000 and the axle ends measured 0.997–0.998. The machine marks could still be seen in the axle ends (the axles holes were reamed) so what I think happened is that the wheels were pressed on (it did take a press to assemble them) originally but were only being held by the peaks in the machine marks. Over time they started to loosen up and as the wheels started to rotate a little bit, it wore the peaks down a little bit causing the diameter of the axle to be a little smaller.

As a temporary fix to the problem, I used a center punch and punched divots into the end of the axles. Around the edges of the divot, the metal would be pushed up a little bit to "increase the diameter" of the axle. This also gives the Loctite something to hold onto other than just a smooth surface. (When doing press fits, the surfaces should be free of things that stick up and could cause galling. I would never use a center punch and make marks on a piece for a press fit.)

In order to drive the wheels back onto the axles, I clamped the left main and left rear wheels to the frame.

***Left:* Figure 121. To help the Loctite hold, I prick punched the axle end. *Right:* Figure 122. In order to press the right wheels back on, I had to back the left wheels to keep the axles from moving. I clamped the wheels to the frame to keep them from moving to the left.**

The rods are not lined up on the left side, but at this stage, that is not a problem. I used Loctite 603 as it is designed to be used in areas where all the oil may not be able to be removed from the surfaces and it can fill up to a 0.003" diametric gap. To drive the wheels back on the axles, I put Loctite on the axle end, lined the key up with the keyway slot on the axle, held a wooden block against the wheel and pounded on the wooden block to push the wheel onto the axle. When the hub was line-in-line with the end of the axle, the wheel was in the correct spot and I then re-installed the set screw so if it does come loose again, it won't come off the axle. Once the wheels and axles were backed, I could easily get the wheels back on the axles.

Figure 123. The R2 and R3 wheels are back on.

The permanent fix for the loose wheels would be to replace the axle or to replace the wheel center. When we made the axles, we make them out of 1018 mild steel so that also gives a third option. The wheel could be pressed off the axle and because the axle is 1018, the end could be welded up (avoiding the key way or the set screw holes) and turned to the new diameter. Only about .004–.005 needs to be added to the diameter to get a good press fit. I don't have a way of doing this, so I used the Loctite 603. A year later, the wheels were still tight on the axles.

One of the things that could have loosened the wheels on the axles is if the engine is out of tram. The centerline from the R1 to R2 wheels should be the same as the L1 to the L2. The same holds true for the R2 to R3 and the L2 to L3 wheels. The crank pins on both sides (all right pins, all left pins) should have the same throw. The side rods should be made to the axle centerline dimensions. To check this, once I got the side rods back on, I rotated the wheels around, checking for binding, particularly when the rods passed over any of the dead centers. At dead centers, the rods and axles are in the same alignment and if a crank pin throw or any side rod/tram dimensions are out, the rods will bind. Past dead center, the wheels and pins have more flexibility. The wheels turned very easily with no binding so I am confident that the problem was specifically the lack of press fit between the axles and the wheel centers.

Figure 124. In a matter of a couple minutes, I got the side rods back on.

The total time for this project was four hours and 40 minutes. It took two hours to disassemble the parts, an hour and a half to machine and install the new pins and an hour and 10 minutes to put it back together and reset the eccentric crank. It would have gone faster if I did not have to remove the set screw from the crank pin.

Figure 125 shows where I could have driven the pin out if the set screw was not holding it to the wheel center.

The 4-6-0 was soon back on her wheels and ready for a test run. These repairs did well, but I did manage to find something else to repair: a piston came off the piston rod. Ah, yes, the never-ending cycle.

Figure 125. The new crank pin is visible between the spring saddle hanger and the front pedestal.

Piston Problems and Lessons in Threading

Around 2002 or 2003, Dad and I were running our former Railroad Supply 2-8-2 at the White Creek Railroad when something happened that brought the engine to an abrupt stop. For some reason the left piston could not reach front dead center. Of course, this did not happen on the ground but on a trestle. I climbed up on the side of the trestle (about seven or eight feet high) and was able to take off the left front cylinder cover (we carry tool boxes with us in case something like this happens. This we learned from experience). Once the cylinder cover had been removed, the problem became quite apparent. The nut that threads on the end of the piston rod (the piston is also threaded) had spun off and jammed between the piston and the cylinder head. We removed the nut from the bore and put the cylinder cover back on. Once we had taken the engine home, we removed the cylinder cover again and loctited the nut back on. We also peened over the end of the piston rods on all of our engines (2-4-0, 2-8-2 and 4-6-0) to keep the nut from backing off.

In April 2012 I finished the track around my house and I got the 4-6-0 in August. Since she had not been run in quite a while, there were some bugs that needed to be worked out. I have a 3 percent grade heading up the front yard and we noticed a pounding or rod knock from the left side when we would head up the grade. Retiming the valves helped some and I removed some slop from the rods. That helped as well, but the pounding still occurred intermittently. In October 2012 some friends were over and the 4-6-0 was pulling five cars with two adults and six kids as passengers around the track. About 40 feet from the top of the grade on the sixth trip the engine went lame. I heard a couple of clunks followed by a "whoosh" and then the engine came to a stop and the left piston could not move beyond front dead center. Needless to say, that ended the ride. I pulled of the left front cylinder cover and discovered that the nut on the piston rod had been driven off. It was a ⁵⁄₁₆–24 nut (½" hex) but it had been squished to about 0.3" wide. When I cracked the throttle, the piston came off the piston rod.

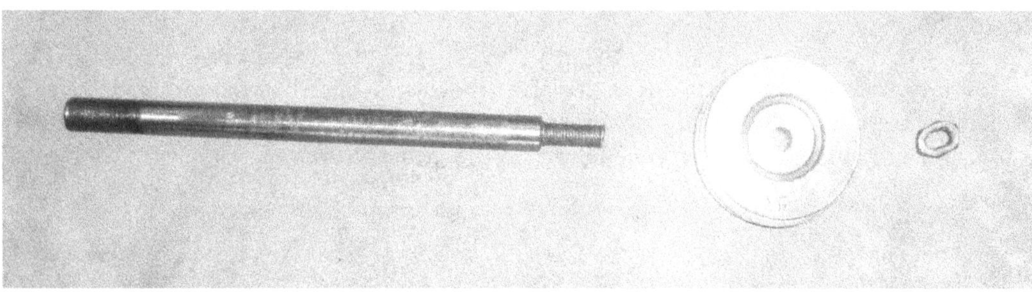

Figure 126. The threads on the left piston rod were cut too small and stripped. The nut came off (½" hex) and the force of the piston coming up to front dead center mashed the nut between the piston and the front cylinder head so the nut was about 0.3" wide.

I inspected the left cylinder bore for any damage and except for some small scrapes and the machine marks from the boring process; nothing else appeared to be damaged.

Upon closer inspection it appeared that the threads in the piston were slightly damaged and not fully stripped. I ran a tap through the threads and they cleaned up decently. When I measured the stripped ⁵⁄₁₆–24 threads on the piston rod, the major diameter was

Above: **Figure 127. The left piston bore did not show any signs of trauma.** ***Below:*** **Figure 128. I made two new rods (one for the left and one for the right as the right side might have also had the threads cut too small) for the 4-6-0.**

0.279". The major diameter should be around 0.308" and minor diameter should be 0.260". It appears that there was not proper engagement between the piston rod and the piston and that the threads on the piston rod were not machined properly. The pounding sound was most likely the piston being loose on the piston rod and pounding against the piston rod and the nut, slowly driving the nut over the peened over end of the piston rod. The "whoosh" I heard was most likely the piston blowing forward off of the piston rod (steam blowing through the hole in the piston) after the nut had been driven off.

I made two new piston rods (in case the right side had the same problem). I had machined and I also bought new 5⁄16–24 jam nuts. I cut the 5⁄16–24 threads about 0.007" large just in case I needed larger threads for

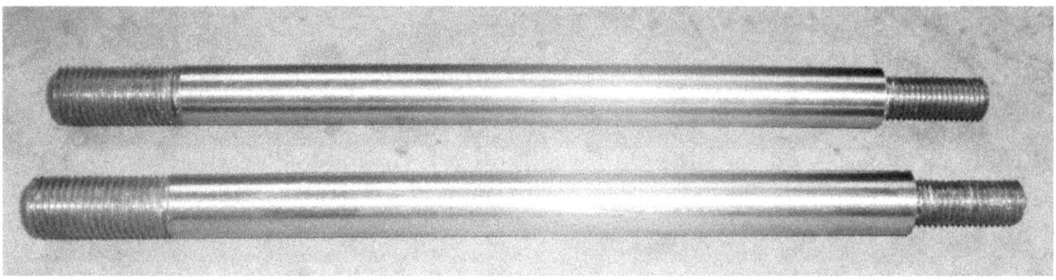

the worn-out piston. I didn't need the larger threads so I finished those threads with a die. The nut and piston had a nice snug fit over the new 5⁄16–24 threads. I loctited the piston and nut on and then peened over the end of the piston rod.

Figure 129. I installed the new left piston rod on the left piston and peened the end over.

Once the left side was reassembled, I took apart the right side to check the right piston rod for any problems with the 5⁄16–24 threads. (Interestingly enough, the 7⁄16–20 threads were fine.) The piston seemed loose on the 5⁄16–24 threads and when I measured the major diameter of the threads they measured 0.304". It seemed like the threads were cut too deep and that the pitch diameter was too small, meaning that the threads did not have proper engagement. When I threaded a new jam nut onto the old piston rod, the nut was loose; however, when I threaded the new nut onto the new rod, the nut had a nice snug fit on the new rod.

Figure 130. The new right piston rod was installed and the end was staked, deforming the threads to prevent the nut from easily coming off.

I installed the old right piston onto the new piston rod and secured the piston and nut using Loctite. To the right of figure 130 is the right piston installed in the cylinder and as the 5⁄16" threaded part was not long enough to peen over, I used a center punch and squished the threads in four places to prevent the nut from being able to unthread itself.

Threading

nominal diameter = size used for general identification (½", ¾", etc.)
P = pitch = 1/(number of threads per inch)
crest width = P/5
root width = P/8
major diameter = diameter across crest = (nominal diameter) – P/11
minor diameter = diameter across root = (nominal diameter) – 1.24P

In the last few years I learned that when cutting external threads there are two choices as to how the threads will be cut. The above calculations are for flat bottom (not a sharp "V") threads. This is the most common as the threads are not engaged all the way down to the root and with a flat root, the minor diameter will be larger than using sharp "V" threads and will give more strength to the bolt, stud, etc. When machining the threads, the threading tool tip will be ground off until the correct length of flat (root width) is reached.

If the threads are cut with a sharp "V" tool, the minor diameter will be smaller while the pitch diameter will remain the same. The minor diameter of a sharp "V" thread is nom. dia. – 1.59P. So the minor diameter of a 5⁄16–24 thread with a flat root is 0.260" and with sharp "V" threads, the minor diameter is 0.245".

When Dad and I built the 4-6-0 between September 2000 and May 2001, we did not know these differences. I think what happened is that we did not know about grinding the tip down to a flat on the threading tool so when we cut to the minor diameter of 0.260" (for the flat root) we used the sharp "V" tool. So this meant that the threads (pitch diameter) were too large (but we did not know why at the time) so other cuts were made until a nut fit over the threads without binding. With those extra cuts, because we did not know why the threads were too large nor by how much they were large, we probably cut the threads (pitch diameter) too small. This would mean that even though the major diameter was close, the minor diameter would have been small and the threads would not have the proper engagement to the piston and it would be a loose fit. This would make sense with the way the threads on the left side failed and left enough thread left in the piston (the minor diameter was only .008" larger than what it should be) that a tap would clean up the threads and still have enough engagement on the new piston rod.

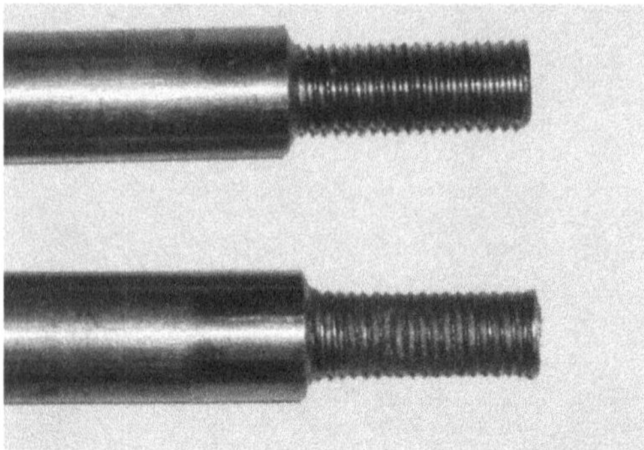

Figure 131. The old rod (bottom) has the stripped out, small threads (with some bronze from the piston in the root of the thread) and the new rod (top) has the fresh threads that are cut to the correct major diameter and pitch diameter.

Figure 131 shows the ends of both the old right (in the top of the picture) and left piston rods. The thread contours look close, but are not quite right.

Centering the Piston in the Cylinder

When I installed both pistons I wanted to have the pistons centered in between the front and rear cylinder heads so there would be even clearance between the piston and the cylinder heads at front and rear dead center. The depth of the bore from the front edge to the rear piston cover was 4.625". The piston was 0.730" and had a stroke of 3.5". The front cylinder head stuck into the bore by 0.156".

1. The space between the cylinder heads will be 4.625 – 0.156 = 4.469"
2. Now subtract out the stroke and thickness of the piston: 4.469 – 3.5 – 0.73 = .239"
3. So there is a total of .239" clearance between the piston and the cylinder heads. To have even clearance on both sides of the piston, .239 will be divided by 2.
4. Even clearance between piston and cylinder head: 0.239/2 = 0.1195"
5. Since the cylinder head sticks in 0.156 from the end of the bore, when the piston is on front dead center, the end of the piston will be 0.1195 + 0.156 = 0.2755" in from the end of the bore. I measured this with my 6" scale. The piston is placed on front dead center and the piston rod nuts are adjusted until it is set the proper distance from the end of the cylinder bore.

Once everything was reassembled, the engine was taken out for a test run. Everything worked well and while there is still a slight rod knock, it is nowhere near as loud and appears to be from slop in the side rods. The total time in taking apart the piston rods, machining the new rods and re-assembly was around three hours. I am also glad that I decided to take apart the right side even though there was not any apparent trouble as I was able to replace that piston rod before it became a problem. Everything is good once again, well, that is until something else needs attention.

Figure 132. The piston is set so it is a shy 9⁄32" from the front of the cylinder casting which will give even clearance to the front and rear cylinder heads.

When There Is No Water in the Glass ... Some Food for Thought

A couple of people have asked me about the two very different approaches to addressing the problem of no water in the sight glass (low water situation). One approach is to dump or kill the fire first, the other approach is to turn on the injectors or water pumps first and then dump or kill the fire. (Of course, neither approach comes close to touching the why of the low water situation in the first place.)

The first approach generally follows the thought that putting cold water into the boiler will cause an overheated crownsheet to fail or that the sudden cooling down of the sheets will cause uneven contractions and/or buckling. However, boiler steel is ductile and does not work harden as other types of steel do. The cooler water is introduced at the front of the boiler, away from the firebox (so it doesn't thermal shock the firebox sheets) and the cooler water will settle to the bottom of the boiler and the hot/boiling water will rise over the crownsheet, so no very cold water will cover the crownsheet. While the overall temperature of the water in the boiler will drop, the temperature drop and low possibility of thermal shock is the lesser of two evils (the other one being a crownsheet failure and explosion). If the crownsheet gets so hot that it is deformed by the water, it most likely has already been damaged by the heat beyond what I would consider the boiler safe to use again.

With oil or propane, the fire can be killed immediately. Dumping a coal fire takes time. On the 2-4-0 it takes about 45–60 seconds to remove the grates, open the ashpan and remove all the hot coals and ash (completely removing the heat source) and this is with a 48 sq. in. grate area. If the crownsheet is bare, it is still being exposed to heat from the fire without any cooling effect from the water. The larger the grate area, the longer it will take to completely dump the fire and remove the heat source. Even if the heat source has been removed, an exposed sheet will still be hot and pliable. Since it is still exposed to the boiler pressure it can fail even after the fire is dumped or killed.

It has been said that if you can't rock the engine and get water to bob up into the sight glass that the fire should be dumped and water should not be added so the firebox sheets aren't damaged by the shock of cool water. Well, if the crown is uncovered and the engine is rocked back and forth, the sloshing of the water still inside can wash over the crown and help dissipate some of the heat, helping to cool the crown as water is added and the fire is dumped.

Some full-size engines had siphons that ran from the throat sheet (below the tubes in the rear tube sheet) to up into the crownsheet. The heating of the water in the siphons would cause the water to rise in the siphon and the water would exit over the crown. In cases of boiler explosions, reports included diagrams that showed where the failure of the crown was and where there was heat damage to the crown. Where the siphons were spilling a thin layer of water over the crown, there was minimal heat damage, even when the water was multiple inches below the crown. The boiler failed where there was no water flow over the crown. I have not yet seen nor heard of a report where the boiler failed because water was added to the boiler. Most of the reports I read had boilers that failed when the water was multiple inches below the crown, giving the crown a decent length of time to overheat.

There have been reports of full-size engines being fired up without water covering the crown. An engine was fired up and got pressure on the boiler with the water level below the top of the crown. The crown overheated and started to buckle. The crown was ruined. Another engine had a fire lit when the water was below the crown, but the boiler was still being filled. The crown lifted as the firebox was heated up, but as the water went over the crown, the crown cooled and settled back down. It loosened up the crown stays but saved the crown.

As a reference for how heat affects the steel, if lifting chains reach 600°F (a dark blue color) the capacity should be reduced to half the original. If lifting chains reach 1,000°F (a grey-red) the chain is condemned. If the firebox sheets are heated up that much due to low water, I won't want to fire the boiler again. The sheets may be damaged, but the damage might not be readily seen.

If the water is ever so low where I can't see it, I will turn on both injectors to try and cool down the crown and then I will dump the fire. I just don't plan on ever having the water get that low.

Revisiting Retubing the Mikado

During the summer of 2002, the steel tubes in the rear flue sheet of our Mikado started to leak. At first it was only a couple, so we borrowed a tube roller and re-rolled the leaking tubes. (The tubes had been rolled into the flue sheets when the boiler was built in 1999.) Once we got the engine home from the Tri-State Locomotive club meet that year, we did a more thorough job of re-rolling. The tubes stayed sealed until that August and we noticed that as we continued to run, we heard the noise of boiling water and noticed water dripping out of the ashpan near the rear flue sheet. I started college a week or so later, so Dad started tearing down the engine to remove the boiler. He decided we should try to retube the boiler ourselves and use copper tubes instead of steel. I helped when I was home for Thanksgiving break and on Christmas break. Just before the end of 2002 the boiler had been retubed, passed its hydro and we had started to reassemble the engine. On January 1, 2003, she steamed for the first time with her new copper flues. Dad wrote an article that appeared in *Live Steam* around 2004. We have not had any problems with leaking flues since.

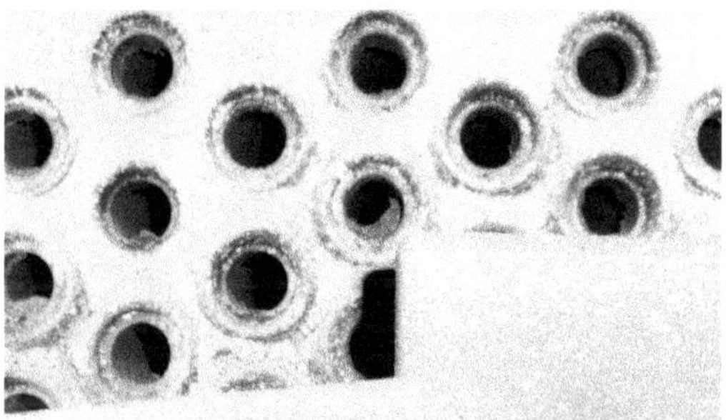

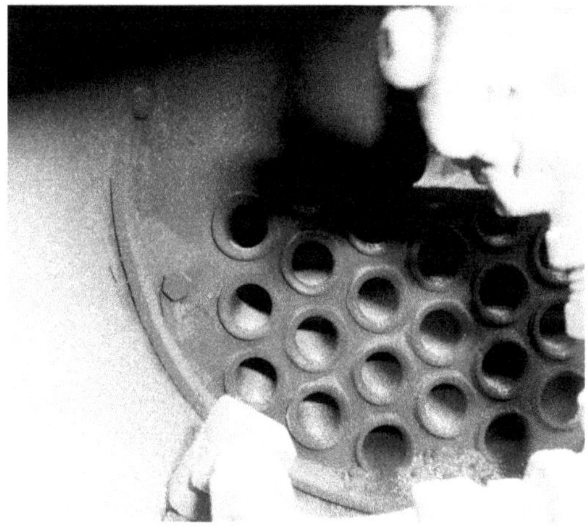

***Left:* Figure 133. The rear tube sheet shows the new tubes that are installed; the old steel tubes were leaking around the rolled joint of the tubes in the tube sheet. *Right:* Figure 134. The front tube sheet shows the new tubes that are installed.**

Figures 133 and 134 were taken just after dumping the fire in July 2010. Figure 133 shows the rear flue sheet (above the arch) and the front flue sheet is seen in figure 134.

Starting in the summer of 2004, Dad and I started working for the City of Traverse City, taking care of and running the 3" scale 4-4-2, The Spirit of Traverse City. In August of 2004 the flues were leaking in the rear tube sheet. The boiler had been reflued around 1997 and the flues had been rolled in and then the ends in the firebox were beaded over against the rear flue sheet. The flues were rerolled and we had no problems for the rest of the 2004 season. However, during the start of the 2005 season, they were leaking again and were rerolled in hopes that it would last long enough for the season. But a couple of weeks later the tubes were leaking (and by leaking I mean waterfalls down the rear flue sheet) yet again and the boiler was retubed. The new steel flues were installed in July 2005 and they were in and beaded over. All was well until the summer of 2007 when they started leaking again. By this time, I was working at the Strasburg Rail Road and I learned why the tubes on our Mikado and the 4-4-2 started leaking only after a few years.

Particularly with 1:8 scale live steam models, steel boilers have copper tubes that are rolled or silver soldered, or they have steel tubes that are rolled or welded. I will talk about using steel tubes in the boilers and why we were experiencing leaking tubes (which still had a lot of life left in them) after only a few years.

Rolled tubes, no copper ferrule. Fig. 1 of drawing 32 shows a cut-away view of a rolled tube in the rear flue sheet. The tube will generally stick out about ⅛"–3/16" from the rear flue sheet. Using a roller, the tube is expanded in the hole in the tube sheet until the steel

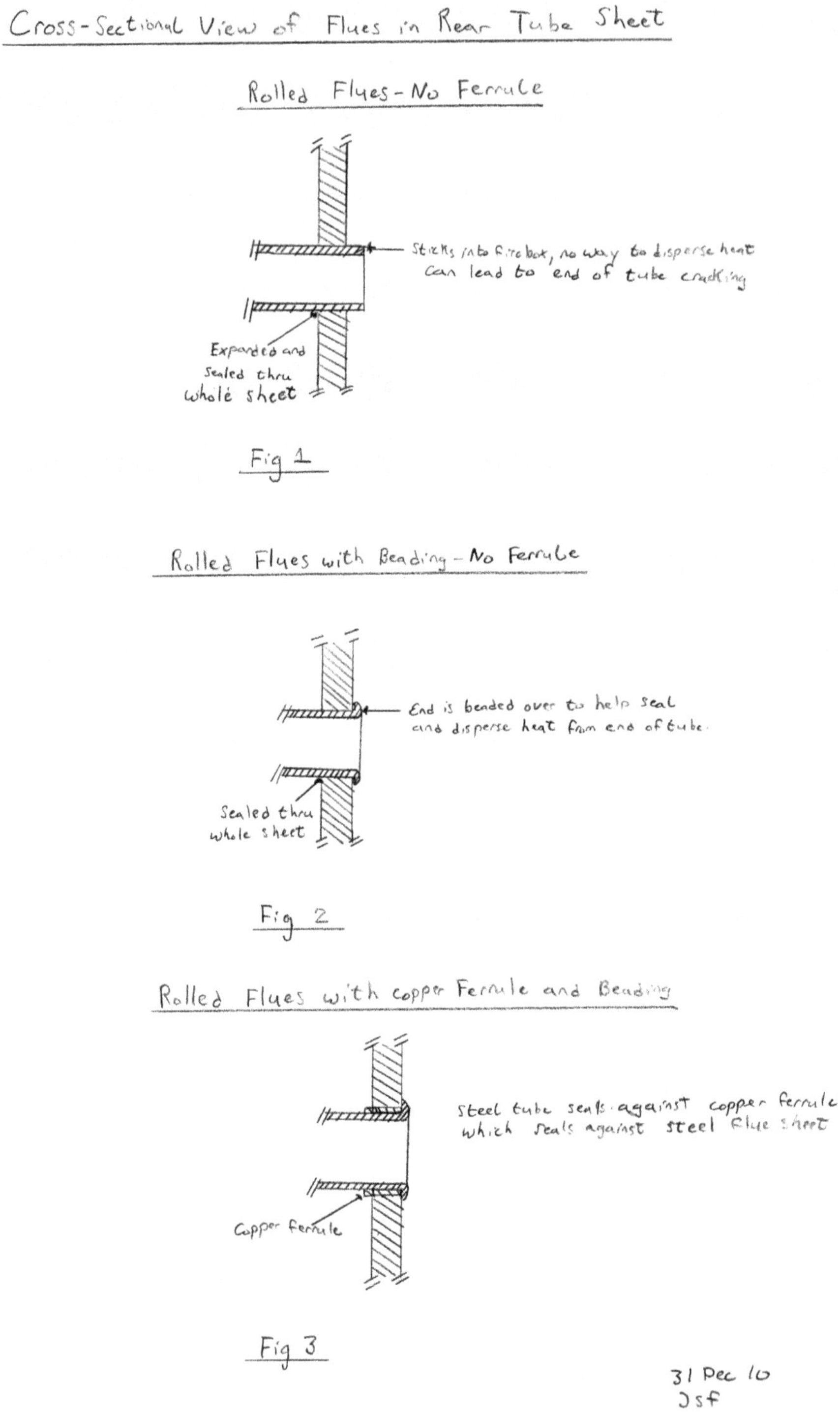

Drawing 32. Figs. 1, 2 and 3 show a cutaway of a rolled tube in the tube sheet (1) that does not have a ferrule and is not beaded; (2) that does not have a ferrule and the end of the tube is beaded; and (3) a tube that has a ferrule applied and has been beaded.

tube is putting enough pressure on the steel tube sheet to create a seal. The sealed joint is across the entire width of the tube sheet. The problem is that with the end of the tube sticking into the firebox, heat from the fire will cause the end of the tube to heat up, sometimes to the point of turning red, and will fatigue the end of the tube as there is no way for the heat to disperse.

Rolled tubes with beading, no copper ferrule. Fig. 2 of drawing 32 shows a cut-away view of a rolled and beaded tube in the rear flue sheet. The end of the tube that sticks out into the firebox will be beaded out until it contacts the rear flue sheet. This is done after the tube has already been rolled. The rolling of the tube creates a seal across the whole width of the rear flue sheet and the beading is supposed to help seal the tube and help disperse the heat from the end of the tube. But because the tube is not joined to the flue sheet, the end of the tube can still overheat and be fatigued. With this and the rolled tubes, the steel tube is creating the seal against the steel flue sheet (steel is not very malleable) and with no way to disperse the heat from the end of the tube, the tubes will generally leak within two or three seasons. This is what we had experienced.

Rolled tubes with copper ferrules and beading. Fig. 3 of drawing 32 shows a cut-away view of a rolled tube using a copper ferrule and beading the end over. The copper ferrule is just a small copper ring that fits over the steel tube where it enters the flue sheet. The copper is more malleable than steel so it can make a better seal as the tubes and flue sheet are expanding and contracting with the temperature differences between the fire and the water.

Dad and I used copper ferrules on the Mikado when we retubed her. Figures 135 and 136 show a tube segment and ferrule side by side and then assembled. The reason why we used the copper ferrule with copper tubes was because we had to make up for the decreased OD from the steel tubes to the copper tubes. If copper tubes are used and rolled in, copper ferrules don't need to be used. When rolling and beading steel tubes, use of the copper ferrule will allow the tubes to continue to stay sealed.

Welded tubes. Fig. 4 of drawing 33 shows a cut-away view of a welded tube in the rear tube sheet. The end of the tube is welded to the flue sheet so heat from the fire can be dispersed from the flue into the flue sheet. The holes in the flue sheet are drilled to have clearance around the tube. The depth of the weld is partially determined by the wall thickness of the tube and rarely does the weld go through the whole flue sheet. This

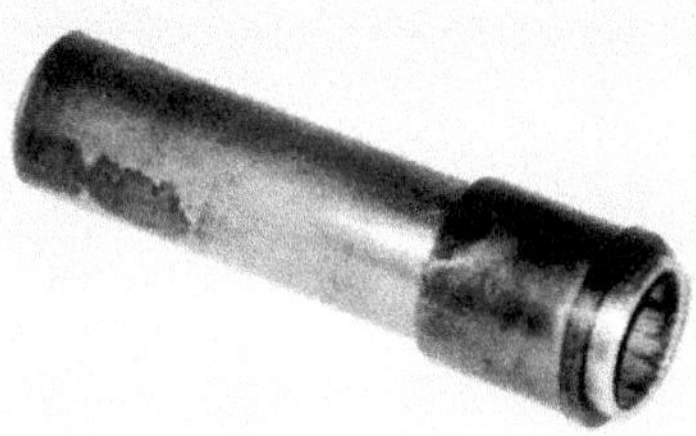

***Left:* Figure 135. Because the copper tube has a thinner wall than the steel tube, a sleeve called a ferrule was made to fit around the tube to take up the extra room.*Right:* Figure 136. The ferrule slips over the tube and when the tube is rolled, the tube and ferrule expand in the hole in the tube sheet.**

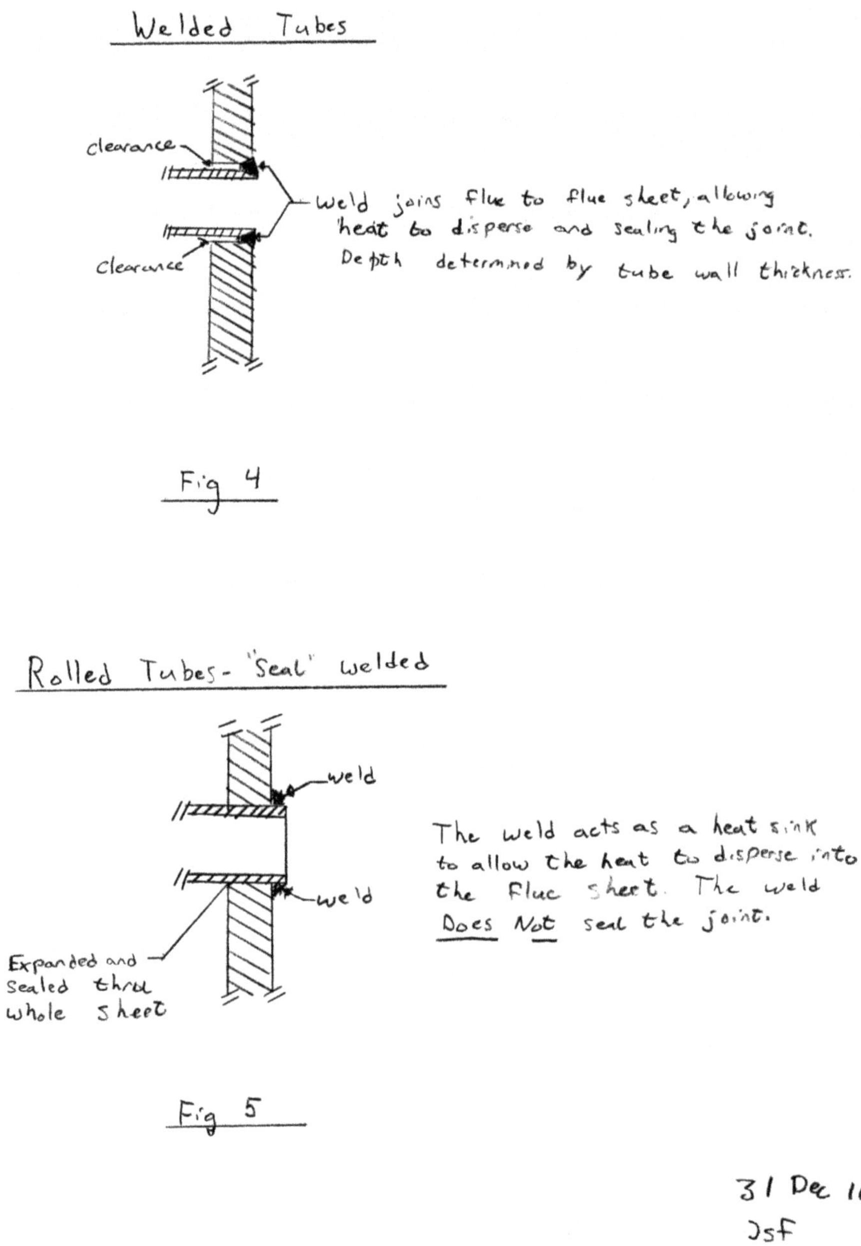

Drawing 33. Fig. 4 shows a tube that is welded in place (the weld seals the tube) and fig. 5 shows a rolled tube (the rolling expands the tube and creates the seal) with a superficial "seal" weld.

leaves clearance between the flue sheet and the tube on the water side and this can cause corrosion of the flue sheet and tubes.

Rolled tubes and "seal" welded. Fig. 5 of drawing 33 shows a cut-away view of a rolled and seal welded tube in the rear tube sheet. The tube is installed with a small part of the tube sticking into the firebox; for 1:8 scale around $3/32''$–$1/8''$ should be good. The tube is then rolled a little. Then the boiler is given a hydro test and the leaking tubes are rolled until they don't leak. This prevents from over-rolling the tubes. Then a bead is welded to the end of the tube in the firebox to join the flue to the rear flue sheet and act as a heat sink to disperse the heat from the tube ends. The name "seal weld" is a misnomer. The weld is not supposed to seal anything, just be a heat sink. The tube should not stick out beyond the edge of the "seal" weld. Once the tubes are welded, the boiler is given another hydro test. Usually some tubes will leak as they were distorted by the welding. Reroll the leaking tubes. If the weld does not adhere to the tube or flue sheet, it does not act as a heat sink and does not pull the heat out from the end of the tube and over time the rolled tube will start to leak.

Firebox end only. Due to temperature differentials in the firebox (the fire can be 2,000°F–2,400°F while the water may be around 375°F), the firebox and flues will expand and contract differently than the front flue sheet, which for the most part has a fairly even temperature. Because of this, the seal welding and copper ferrules only need to be used in the rear tube sheet. The tubes in the front flue sheet can be rolled in normally.

Figure 137 shows the rear tube sheet of the Spirit of Traverse City shortly after the seal welding was completed. The tubes were installed in July 2005 and the welding was done around July 2007. The rust that is seen is from the tubes leaking under the hydro before they were rerolled.

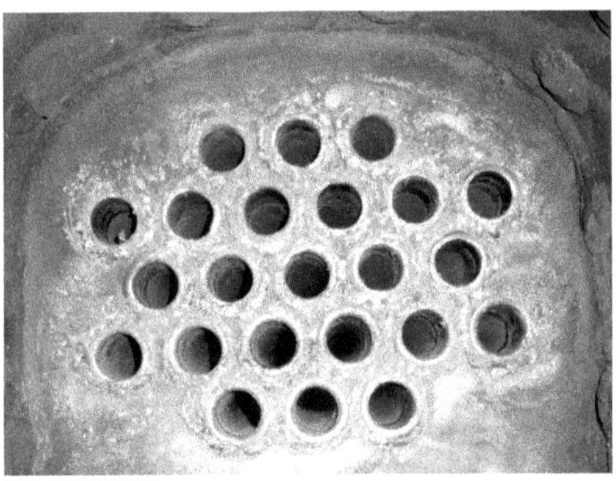

***Left:* Figure 137. The tubes on the Spirit of Traverse City were simply rolled and beaded and started to leak after a few years. They were then seal welded and rerolled. *Right:* Figure 138. The tubes on this standard gauge engine were rolled and seal welded.**

The seal welds in figure 138 can easily be seen around the tubes on the rear flue sheet on this full-scale engine.

When all of the problems with leaking tubes started, Dad and I didn't know why they started leaking after so few years of service. We have not had any problems with leaks after we retubed the Mikado and she is still steaming freely. Working at the Strasburg Rail Road I have learned why the problems occurred and what to do to prevent it; this also helped to keep the Spirit of Traverse City running.

Making an 0-4-0 into a 2-4-0

Around 2001 or 2002 I decided to try and add a lead truck to the Railroad Supply 0-4-0 that my dad and I had built. The engine was started in 1990 and in 1996 she ran for the first time. Figure 93 shows her being run by my grandfather at the White Creek Railroad in the summer of 1996. Over the first several years we had several derailments where the engine would pick a point of a switch or pick a frog and would derail and damage the cylinder cocks.

In order to prevent more damage to the cylinder cocks I attached guards to the locomotive frame which extended under the cylinder cocks and added a heavy-duty step board on the front bumper that would ride along the rail head in the event of a derailment.

Figure 139. The cylinder block on the 0-4-0 sits low to the rail head so I added a guard under the cylinder cocks so the guard takes the brunt of the force in a derailment. The guard was modified slightly when the lead truck was added.

Around the summer of 2001, Dad and I were running the 0-4-0 and 2-8-2 at the White Creek Railroad. I was running the 0-4-0 in front of Dad, who was running the 2-8-2, and the 0-4-0 decided to jump off the rails at a switch to a passing siding. The front bumper bent, the guards for the cylinder cocks were basically destroyed and three out of the four cylinder cocks were damaged beyond use. Once I got the engine back on the rails, we disconnected the engine and tender, pushed the engine to a passing siding and the 2-8-2 coupled on to the rear of my train. We picked the 0-4-0 up and set it on a flat car that was in the consist and brought her back to the steaming bays on a flat car.

Once the engine was back in our shop, I started to figure out how I wanted to make a lead truck for her. Looking at several other engines (model and full-size) it appeared that the lead truck wheel was around half the diameter of the driving wheel. The drivers are 6" so I chose a 3" pilot truck wheel, which were turned from a piece of 3.5" diameter piece of steel. The axle was machined out of a ¾" diameter steel bar, and ⅝" ID × 1⅜" OD single-row, double-sealed ball bearings were used in the journal boxes.

I took off the front bumper of the engine and saw that there was a limited amount of clearance between the frame under the cylinder block and the rail head. Instead of having a pivot bracket where the stud pointed down, I mounted the bracket so the pivot stud pointed up, leaving enough clearance between the end of the stud and cylinder block to slip the truck frame over it.

Figure 140. The pivot for the lead truck is under the cylinder block and due to the lack of clearance below the frame, the pivot pin sticks up.

In the top of the stud, I drilled a ³⁄₃₂" hole for a homemade, longer than normal clevis pin. In the event of a derailment where the lead truck is damaged, I can use the poker for the fire and pull out the clevis and remove the lead truck and run her as an 0-4-0.

After mounting the pivot bracket, I took some rough measurements to give me an idea of how far in front of the frame the pilot wheels should be to (1) clear the frame when going around curves and (2) to not look out of proportion.

Top left: **Figure 141. The rear view of the lead truck frame.** ***Top, right:*** **Figure 142. The centering device (using extension springs to keep the roller centered) is easily seen. New, stronger springs were installed in 2015 to help center the lead truck. The old springs were too weak; when the lead truck was installed I did not know how strong they needed to be.** ***Bottom left:*** **Figure 143. The frame is made out of ½" thick steel with a ¼" plate used as the floor for the roller. The roller block can be seen sticking above the frame.** ***Bottom right:*** **Figure 144. Around 2012 I added another brace that is seen in the center of the truck.**

Figure 145. The journal box is made out of steel and holds a single-row, double-sealed ball bearing. The pedestals are ½" square stock.

Figures 141–144 show the truck frame that I built using those rough dimensions. The pedestals were made out of ½" square bar stock. The main frame was made out of ½" square stock and ½" × 1¼" stock that was lying around. The pedestal binders were made from ⅛" × ½" stock, the pivot was made from ¼" × ¾" stock and the braces from the pivot to the main frame were made from ⅛" × ½" and ⅛" × ¾" stock. In order for the bracing to clear the frame on tight curves, I had to trim the braces near the pilot truck main frame.

Figure 145 shows the journal box in the frame. I have a coil spring in between the top of the frame and top of the box. The spring sits in recesses in the box and in the frame which holds the spring centered between the pedestals and centered over the ball bearing.

Rather than using heart links as a centering device, we followed the suggestion of a fellow live steamer, who recommended that we use something that rolls along a plate and use springs to keep it centered. The centering device can be seen in the figure 142. It is made out of 1" square aluminum and uses two 5⁄16" diameter rollers that roll along a ¼" thick plate which is welded to the bottom of the pilot truck frame. The two modified extension springs keep it centered. The top of the centering block has been milled to accept the hex of a ½-13 bolt, which is used as the centering post. (Note: in 2016 I replaced the extension springs with stronger ones and that really helped center the lead truck. When we were first doing this, we were afraid to use a stronger spring and have it not be flexible enough. That was not the case. It was too weak and didn't do a good job of having the lead truck lead the engine. The lead truck would not track right and would kick to the left when running forward and pull to the right when running backwards. I returned the lead truck wheels to the correct profile and same size as they were slightly worn. With the engine centered, the pivot was off 1⁄32" from center so I replaced that but that did not fix the problem. The stronger springs did and the lead truck tracks centered when running on straight track and guides the engine better. The engine is also more stable and doesn't nose as much since the stronger springs were installed.)

Once the frame had been made up and the axle set installed, I attached the lead truck to the pivot on the frame and then figured out how to extend the front of the frame. All of the metal for the frame extension is ⅜" thick bar stock.

Figure 146. I had to extend the frame to fit the lead truck. Since I was reusing the front bumper (which fits inside the frame) I had to make a stepped extension to fit inside the existing frame and then widen it to fit around the bumper brackets.

After I got the parts cut and in place, I found out that I needed to move the bumper forward a little bit more which is why not all the holes on the frame extension line up with the holes in the bumper. After the frame was complete, I made the deck plate (out of ⅛" thick plate that was lying around) and bolted it to the bumper and to the frame extension with 10–32 socket head cap screws.

***Above:* Figure 147. The center pin has a coil spring that bears against the pilot deck. The centering device fits over the bolt head and the coil spring transfers some of the locomotive weight to the lead truck. *Below, left:* Figure 148. The lead truck is installed and the center pin bolt head is fitting into the centering device. *Below, right:* Figure 149. The center pin sticks through the pilot deck and has a nut installed on it to keep the pin from falling out when picking up the engine and/or removing the truck. A 1" square piece of aluminum was installed on the deck (which is ⅛" thick) to act as a guide for the center pin.**

Figure 147 shows the bottom of the deck plate and the centering post. The centering post is a ½-13 bolt and a spring is held between the bolt head and the deck of the locomotive. The weight of the front will be transferred through the spring and bolt onto the lead truck.

Figure 149 shows the top side of the deck plate and the centering post. The 1" square aluminum block is added to give the centering post more stability so it won't rock as the engine goes around a curve. The ⅛" thick plate does not have enough surface area to keep the post from rocking side to side when going around curves. The smokebox/deck supports were added to help improve the look and I added supports to the ends of the running boards.

The way I have the truck sprung is not prototypical. A two-wheel lead truck is usually equalized with at least the first driving axle to transfer weight and force back and forth when traveling over the track to keep the other wheels solidly on the rail when one wheel is picked up. As seen in figure 140, there was very minimal room underneath the cylinder block to try and install an equalizer from the lead driving axle up to lead truck. Also, since I was unsure how successful the truck would be, I didn't want to take the time to tear apart the spring rigging to have it equalized to the lead truck only to find out the truck was a failure and then need to change it all back again. Because the lead truck is not equalized with the drivers, if the lead truck is damaged, I can remove it from the locomotive without worrying about messing up the spring rigging and can run it as an 0-4-0 to get it back to a steaming bay.

Figure 150. To keep the truck from falling off when rerailing the engine or picking the engine up, a bracket was installed to the bottom of the bumper (the loose bolt has been retightened) to extend under the truck frame to limit how far down the truck frame can move in relationship to the locomotive frame.

In order to prevent the lead truck from dropping down too far when the engine is being picked up (either onto a stand or being put back on the track) I added a ⅛" × ¾" piece of steel to the front bumper to minimize how far the truck frame can fall, seen in figure 150. (Note, the loose bolt that holds it to the bumper has since been tightened.)

The lead truck has worked out well. There is enough weight being taken by the lead truck so that the wheels don't pop off the track with every little bounce, but not enough where it really takes away from the weight on the drivers. I have noticed that the lead truck helps to guide the engine into switches and curves and in a couple of derailments, the lead truck was damaged (and was removed to run the engine back to the steaming bay) and took the brunt of the force while the cylinder cocks were not damaged. In 2012 I added a brace on the bottom of the truck frame leading from the pivot piece to the front of the truck frame to help add extra strength and not have it bend as much in a derailment. I have been very pleased with how the lead truck works and that it does not look out of place on the engine.

Equalizing

A fellow live steamer asked me about equalizing and spring rigging. He said he is building a miniature steam engine (I forget what wheel arrangement) and he asked me how the springs worked on the full-size. He said he was told that the full-size equalizing doesn't work in miniature as the springs are always working against each other as the engine moves down the track (which I told him is what they are supposed to do) and he was told to just use coil springs between the top of the boxes and the frame and to not bother machining all of those spring rigging parts as they "don't work anyways." He said it was cheaper and easier to just use coil springs. (I strongly encouraged him to use the prototype equalizing as it does work well in miniature for the same reason it was developed for the full-size.)

The purpose of equalizing is to transfer force/weight from one set of drivers to another so if one set is picked up, more force is transmitted to the remaining drivers to keep them on the track. As a lead truck swings, more weight is transmitted to the drivers to keep them on the track and helps to guide the engine through curves and the lead truck wants to swing back to its neutral position with the least amount of force being transmitted back to the drivers. There is much more to it than just keeping the engine sitting level when standing still on level track. With coil springs in between the top of the box and the frame, the movement of the driving box is directly imparted on the frame; with equalizing it is minimized.

In order to keep the frame from getting the direct shocks and movement of the driving boxes, full-size steam engines use a system of levers connecting the driver springs to the frame and to each other. As the driving box moves up and down, the spring above it will also move up and down. This movement is transmitted through the spring hangers to the equalizers (a lever) which reverse the movement and apply the reversed movement to the next spring. The fulcrum of the equalizer is what directs the force to the frame. Because of all the equalizers (levers) the effect on the frame is minimized as the boxes move up and down in the frame. If one wheel is picked up ½", the frame might be picked up ⅛" or less.

In the U.S., steam engines generally had a three-point suspension. If an engine did not have a lead truck, generally the lead driving axle had a cross equalizer which connected the front left spring to the front right spring. The lead axle would set the height of the front of the engine. The following drivers would be equalized on each side; the right rear drivers would be equalized with each other to set the height of the right side of the frame and the left rear drivers would be equalized with each other and set the height of the left side of the frame. On straight and level track, the front corners of the frame should be the same height and the rear corners should be the same height. If the top of the frame is flat with no height differences, the four corners should be the same or quite close. The springs and equalizers should also be level (the ends of the springs are at the same height). The driving boxes should be about halfway between the pedestal binder and the top frame rail so as the box moves up and down it won't bang into the frame or binder.

Since track is not perfectly smooth, the springs will be rocking back and forth on the springs saddles (and they also act as equalizers) as well as having the equalizers rock back and forth as the engine runs down the track and the wheels pick up some or drop some.

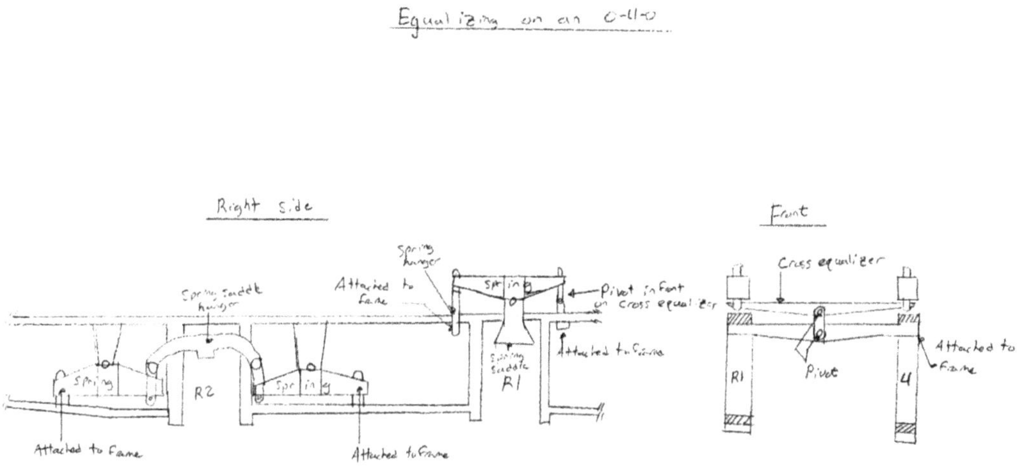

Drawing 34. With an 0-4-0, the springs on the front driving axles are cross-equalized with each other and the rear driving boxes are spring separately. The pivot point of the cross equalizer is the front point of suspension and the springs for the rear driving boxes give the rear two points of suspension.

For an 0-4-0, the lead driving axle is cross equalized by a bar (cross equalizer) so if the R1 driver is picked up, more force will be transferred to the L1 to keep it on the track. The rear drivers were equalized separately to give the three-point suspension.

If an engine has a two-wheel lead truck, the lead truck is generally tied in with the front driver(s). If an engine has a four-wheel lead truck, the older designs did not have the lead truck tied in with the drivers, but the lead truck set the height of the front of the engine and the right-side drivers were equalized together and the left side were equalized together. Pennsy (the Pennsylvania Railroad) did start equalizing the four-wheel lead trucks with the front driver(s). The trailing truck wheels are generally equalized with the rear driving wheels (right trailing with right rear drivers, etc.).

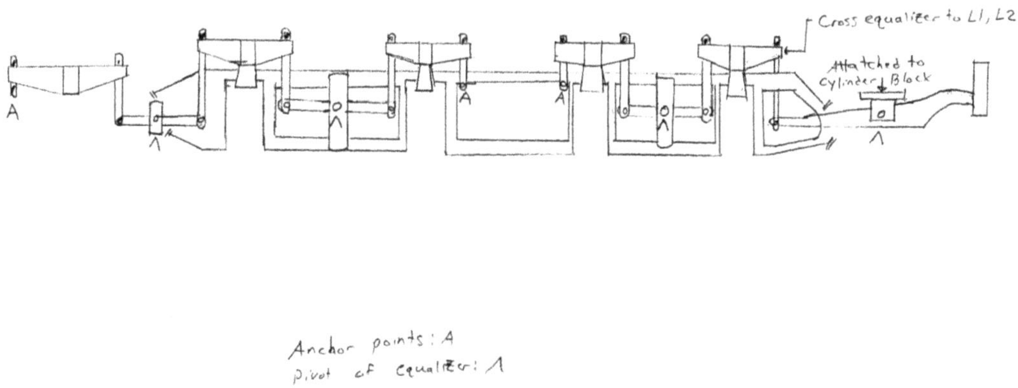

Drawing 35. The lead truck is tied into the front two driving wheel sets through the beam heading back to the cross equalizer. The front driver springs are tied into the second driver springs through equalizers, which is a lever. The right two rear drivers are equalized together and with the right trailing truck wheel. The left two rear drivers are equalized together with the left trailing truck wheel.

On a 2-8-2, the force from the springs is applied to the frame at the fulcrums of the equalizers between the first and second axle sets, the fulcrums between the third and fourth axles, the pillow blocks where the springs are tied directly to the frame between the second and third axles, the fulcrum of the equalizer to the lead truck and the rockers on the rear of the trailing truck. On drawing 35, pivots of the equalizers are marked with an upside down "V" and anchor points where the springs are anchored to the frame are marked with an "A." There is a cross equalizer connecting the Rl, R2 and the lead truck to Ll and L2; the cross equalizer is similar to what it would be on an 0-4-0.

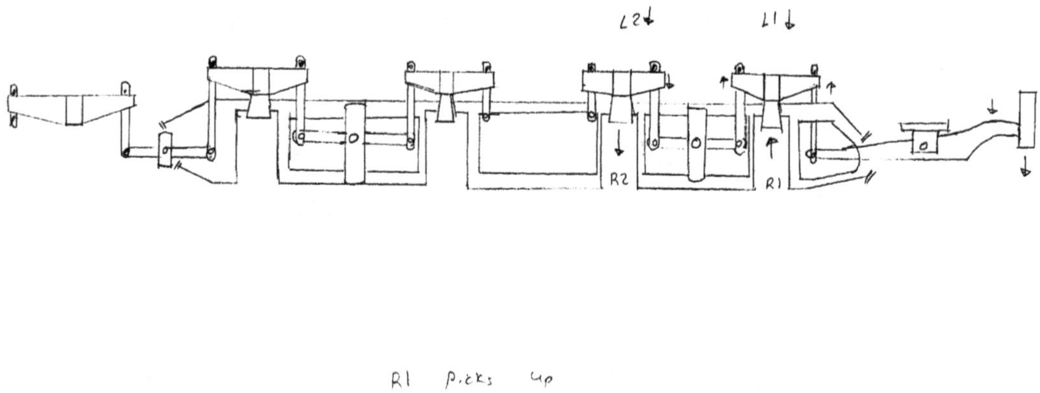

Drawing 36. With the equalized system, if one wheel is picked up, the series of levers and springs transfers force to the other equalized wheels to keep them on the track. In this case, the Rl driver is picked up. This puts more force on the R2 driver (through the equalizer) and the lead truck (through the cross equalizer and beam to the front truck) and the Ll driver (through the cross equalizer) and the L2 driver (through the cross equalizer and equalizer between the Ll and L2 wheels).

If the Rl wheel moves up, an upward force will be applied to the fulcrum between the Rl and R2 drivers, the fulcrum to the lead truck and the fulcrum between the Ll and L2 drivers. However, an additional downward force will be applied to the lead truck, R2, Ll and L2 drivers to keep them from picking up off the track.

When the engine starts around a curve, the lead truck will swing off to the side. Due to the rockers or swing links, more weight will be applied to the lead truck (the least amount of force is applied on straight track when the lead truck is straight) and the bissel post will be picked up and will help guide the frame (as it is captured and rides in the front frame member) into and around the curve. As the swing links/rockers pick up the lead truck end of the equalizer, that will apply a downward force onto the lead driver's springs and put more force onto the front drivers to keep them from wanting to climb the rail. When the trailing truck moves off of center, the rockers will force the truck frame down, putting more weight on the springs. As the rear of the trailing truck spring is forced down, the front of the spring will be forced up by the trailing truck box (which will keep the center at pretty much the same height off the rail). This will put more force on rear drivers, keeping them from climbing the rail.

When the bissel post lifts as the engine goes around the curve, more force will be transferred back to the Rl, R2, Ll and L2 drivers through the equalizers and cross equalizer. When more weight is forced onto the trailing truck frame as it swings around a

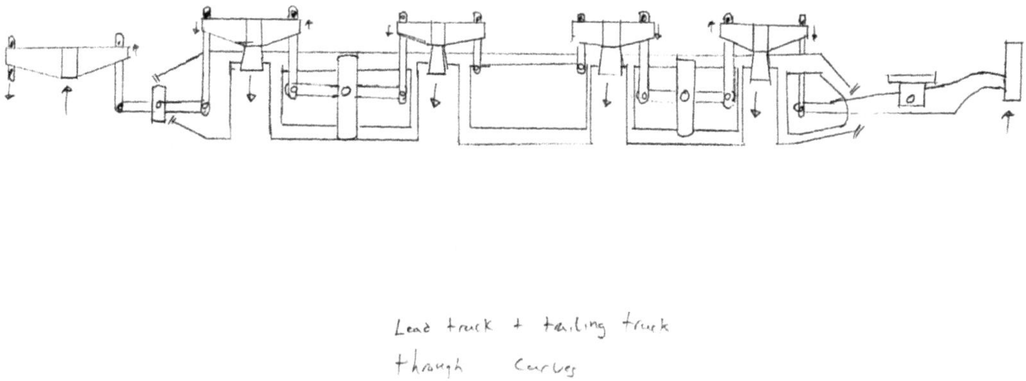

Drawing 37. When the 2-8-2 heads through a curve, the heart links on the lead truck will pick up the equal beam and transfer more force to the first two driving axles. The rockers on the rear truck will put more weight on the trailing truck which will transfer more force to the right rear and left rear drivers through the equalizers.

curve, the right trailing truck spring will transfer more force to the R3 and R4 drivers through the equalizers and the left trailing truck spring will transfer more force to the L3 and L4 drivers.

When the engine is on straight track, the rockers on the trailing truck and rockers/heart links in the lead truck will be exerting the minimum force on the equalizing. The equalizing will want to come back to the minimum force state, which is what helps guide an engine through curves.

If an engine is designed with coil or die springs between the top of the driving box and the frame, the frame will be taking the direct impact from the driving boxes. If one driver picks up, it can pick up that side of the frame and lessen the weight on the other drivers which can make them more likely to climb the rail and derail.

Lubrication for the Cylinders

The pistons and valves need a good, proper supply of lubrication. The oil that is used should be steam oil or a heavy-duty oil that is designed to take the high temperatures (125 psi is around 350°F) of steam lubricators that will pump the oil into the steam lines. The 2-4-0 has a single-feed lubricator from Coles. The 4-6-0 has a single-feed lubricator and the 2-8-2 has a dual-feed lubricator, both from American Model Engineering Supply. The single-feed lubricators feed the steam oil into the steam line, just before it branches to the two cylinders. The dual-feed feeds oil to the steam just before it enters the valve chamber. All of the feeds have a check valve before the oil enters the steam line to keep steam from blowing back to the lubricator.

The feed of lubrication should not be so much that it drains the tank quickly and causes oil droplets to come out the stack, but enough that there is a fine film of oil on the piston and valve rods and on the bores. If too little oil is fed, the bores may be dry which can lead to scoring and extra friction that the engine needs to overcome.

Figure 127 is of the left piston bore on the 4-6-0 and I took that picture in the fall of 2012. She was finished in 2001 and the bores are not scored and the machine marks are still visible.

Top: Figure 151. The 2-4-0 uses a single-feed lubricator from Coles Power Models. *Bottom:* Figure 152. The 4-6-0 uses a single-feed lubricator from American Model Engineering.

Figure 153. The 2-8-2 uses a dual-feed lubricator from American Model Engineering.

When Dad and I started working on the Spirit of Traverse City (15" gauge 4-4-2) in 2004, she did not have a lubricator for the cylinders. Previous operators pumped three pumps of 10W-30 oil into the valve chamber each time they stopped at the station after a 0.4 mile trip. They also had to hone the bores each year due to scoring and replaced the rings every few years as they started leaking.

Figure 154. The Spirit of Traverse City uses a dual-feed lubricator from American Model Engineering.

We installed a dual-feed lubricator (from American Model Engineering) in July 2004 and it has operated well from 2004 until 2011 when the City of Traverse City got rid of the train. The engine ran an average of 35–40 miles per day for about 100 days a year and we have never had to replace any part of the lubricator.

Since adding the lubricator the engine drifted a lot easier and the scoring subsided. During the winter of 2004–05 we honed the bores and installed new rings, one new piston and two new valves. We did not need to touch them after we did that.

While some people feel that the water (condensation) in the steam will lubricate the cylinders, it really won't. I know a few people who had compound engines on steamboats and they said when they installed a lubricator for the cylinders that the engines ran smoother, quieter and did not have to work as hard and they were able to get more speed once they had proper lubrication.

All of our mechanical lubricators are driven off the link so they will receive a constant supply of oil. If a lubricator is driven off the top of the lead and lap lever, when the engine is hooked up, the oil feed will be slower than when the engine is in the corner.

Note: As of 2017, American Model Engineering and Coles Power Models no longer exist. Some of their lubricators may still be floating around. LocoParts sells lubricators.

Throw, Quarter, Side Rods and Such

The throw of the crank pin is half the stroke of the piston; it is the distance from the centerline of the axle to the centerline of the crank pin. Quarter refers to angle offset from the crank pins on the right side of the engine to the crank pins on the left side. In standard practice, when an engine has two cylinders, the crank pins are offset by 90 degrees. Since the driving wheels are connected by steel side rods, the quarter on all the driving wheels needs to be the same and the throw of the crank pins needs to be the same on each side of the engine. The throw doesn't have to be the exact same from the right side to the left side, but it needs to be within tolerance of what the piston can handle so the piston doesn't slam into the cylinder heads. The driving axles should be set so the centerlines between the right driving wheels are the same as what they are between the left driving wheels as well as being square to the frame (no parallelograms). The side rods should be machined to the driving axle centerlines so everything is in tram. If the rod connecting R1 to R2 needs to be 0.015" longer than the rod from L1 to L2, then the wheels are out of tram.

If the rods bind when crank pins are on the axle centerline, one of the crank pin throws may be off. If a side rod is acting too long when the crank pin is behind the driving axle, but too short when the crank pin is in front of the driving axle, the throw on that crank pin is too short.

If the quarter is off, the rods will more than likely bind when the crank pins pass by the 7:30, 10:30, 1:30 and 4:30 positions. This is when both sets of rods are not on the axle centerline.

The bushings should not be tight on the crank pins nor should there be any sharp edges in the bore. Sharp edges (scores, from the oil/grease hole, etc.) will wipe away the lubrication. I prefer to have around 0.003"–0.005" clearance when the bushings are new. When running in one direction, the rods will bear on the same side of the crank pin and over time will wear that side of the crank pin. When running in forward, when the crank pin is at front dead center, the rod will push on the outside edge of the crank pin to turn the wheel. When the crank pin reaches rear dead center, the rod will pull on the outside edge of the crank pin to turn the wheel. This type of wear will wear the crank pin out of round and make it not true to the center of the crank pin. If an engine operates more or less equally in forwards and reverse, the crank pin will wear more evenly, but still won't wear perfectly round.

When more than two driving axles are used, a knuckle pin is used in the side rods. The knuckle pin gives the rods vertical flexibility that solid rods don't have. As the wheels pick up and down in the frame, the rods need to have flexibility so the engine does not tear itself apart and runs more smoothly.

Due to the angularity of the main rod, the piston does not reach its mid-stroke when the crank pin is at the 12 or 6 o'clock positions. The movement of the piston is not a linear relationship to the movement of the crank pin due to the angles that the main rod moves through. The mid-stroke is measured by marking a dead center (let's say the front dead center) and then moving the piston/crosshead half of the stroke. The crank pin position for the piston mid stroke will be past the 12 o'clock position and before the 6 o'clock position (like 12:45 and 5:15 positions).

Figure 155. With the Alligator crosshead, when the engine is running forwards, due to the angulation of the main rod, the force will be applied to the top shoe and crosshead guide. When running in reverse, the force will be applied to the bottom shoe and crosshead guide.

Figure 155 is of an Alligator type crosshead. When the engine is running forward, the crosshead will bear against the top crosshead guide due to the angularity of the main rod. When running forward in this picture, the piston is being pulled forward. The main crank is above the axle so through the main rod the crosshead is being pulled up. When the crank pin is below the axle centerline and the engine is running forward, the piston will be pushed backwards and through the angle of the main rod will be pushed up.

This crosshead has bronze shoes that bear on the crosshead guides. Many models have a single-piece cast crosshead. We made a steel center section of the crosshead and bronze shoes, in similar fashion to the full-size, so when the shoes wear out, only the shoes need to be replaced and not the entire crosshead.

Wheels and Trucks

The IBLS standards for wheel dimensions give a very basic idea for the profile of the wheels. The wheels for 1:8 scale should be ¾" minimum in width; wheels wider than

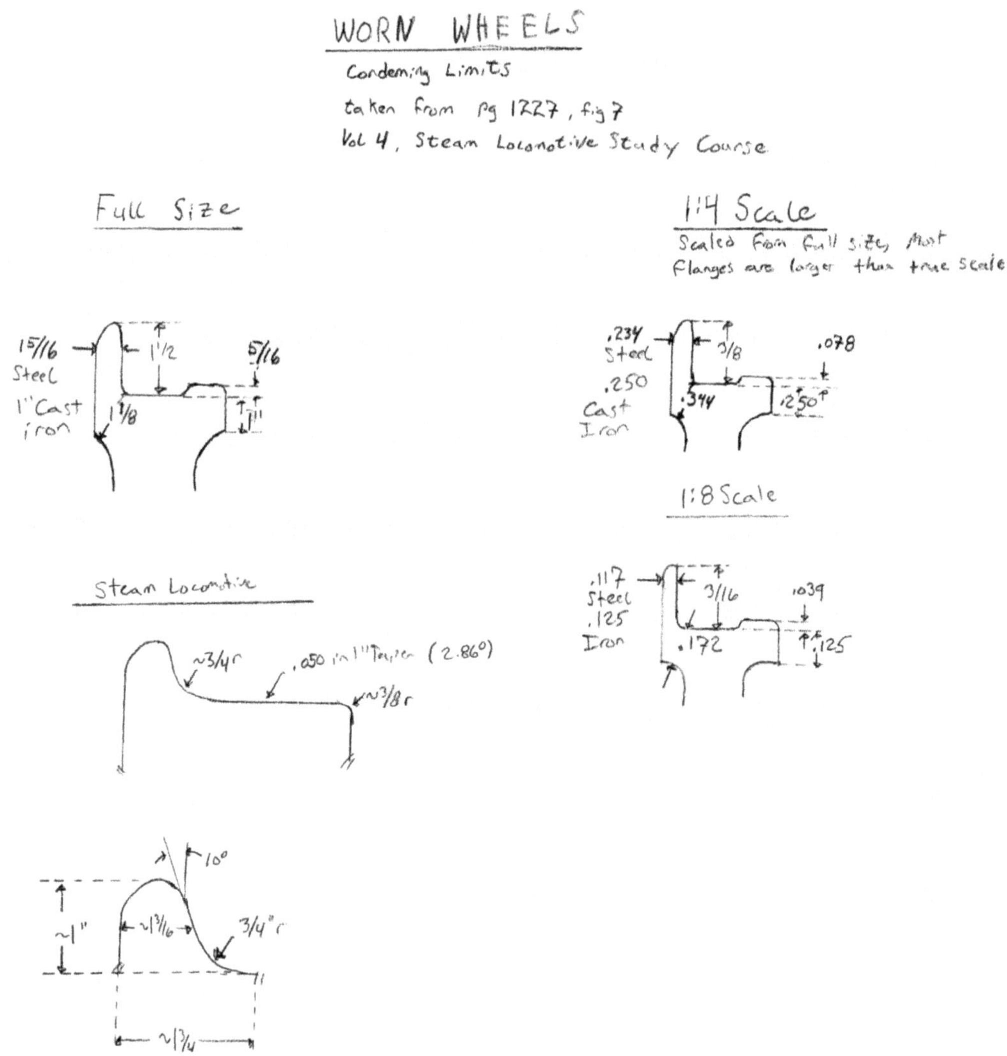

Drawing 38. The full scale/FRA condemning dimensions are listed for wheel sets and are scaled down to 1:4 and 1:8 for reference.

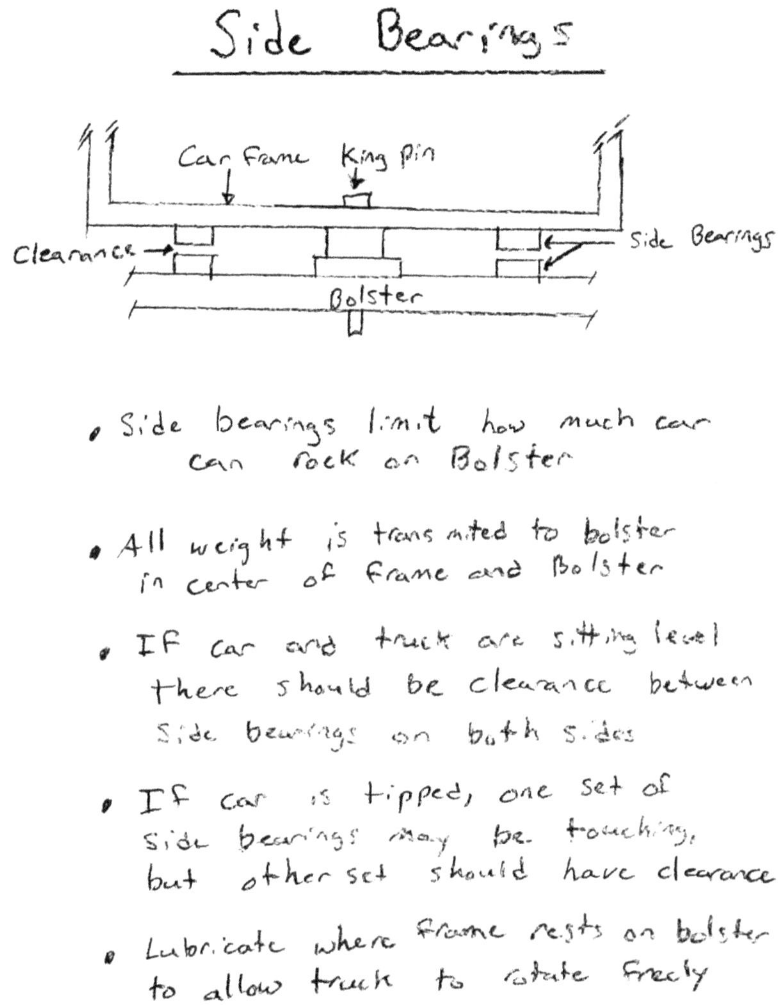

Drawing 39. The center plate rides on the bolster and there is clearance between the two side bearings when the car is level on straight, level track.

that do fit in the IBLS standards as there is no maximum width given. Narrow gauge equipment on 2.5"–3" scale can run on 1:8 scale track, but their wheel widths will be wider than ¾". For a 1:8 scale flange, a maximum height of 3⁄16" is given; a flange shorter than this fits in with the IBLS standard and if it has the proper profile, it can work very well. The wheels have a taper and are a larger diameter at the base of the flange/fillet. This taper will help the axle sets find their home when running down the track; they will hunt side to side and the flanges shouldn't really touch the rails on straight track as the tapers are looking for their home where both wheels are turning on the same diameter. On a curve, the axle set will drift to the outside of the curve. This puts the larger diameter of the wheel on the outside rail with having the smaller diameter of the inside wheel on the inside rail. The outside wheel will need to travel more than the inside wheel through the curve and this difference in diameter helps the wheels go through the curve without having one wheel slide or the other wheel skid as much.

The back-to-back dimensions, the profiles and the widths of the flanges are critical. If the back-to-back is too wide, the axle set can interfere and pick points on frogs. If the back-to-back is too narrow the axle set can interfere with guard rails at the frogs. If the flanges are sharp, they can cut into the rails and quickly ruin the rails. Also, sharp flanges have a greater risk of climbing the rail and derailing.

IBLS does not list any condemning limits on wheels. I have taken the full-size condemning limits for wheels and have scaled them down as a reference for knowing when to take the wheels out of service and/or put new tires on.

Trucks should be rigid and yet flexible to ride down the track. The bolster should hold the truck frames rigid to keep the axles in tram (axles are held perpendicular to the gauge and the front wheels are the same distance away from the rear wheels. One side frame should not be able to be held still while the other side frame can move forward and backwards) and yet allow one wheel to be picked up a fair amount and keep the rest of the wheels on the track. With 1:8 scale equipment, I like to be able to pick up one wheel ¼" with the other three remaining on the rail.

The typical practice is for the car weight to rest on the center of the bolster and have gaps on the side bearings which are closer to the ends of the bolster. This gives minimum surface area and friction for turning. If the car is also bearing on both side bearings, the surface area and force required to have the truck turn are greatly increased. In theory there is an equal clearance between the side bearings. If the car is tipped, one bearing may have minimal clearance or is touching, but the other side bearing should have clearance. It is a good idea to periodically lubricate (graphite works well, whether by itself or mixed with some oil or grease) where the bolster turns on the car frame. If a car rocks too much, there may be too much clearance on the side bearings (I have just under 1/32" on m 1:8 scale) or the springs are too weak and allow the bolsters to rock in the side frames.

Basic Lights

The prototype engine has a headlight on the front, one on the tender, classification lights on the front of the locomotive, marker lights on the tender, lights for the steam pressure gauge, brake pressure gauges, water gauges, a light for the gangway and several lights on the running boards for ground work on the locomotive. The lights that I have found that I need on scale engines are the headlight, lights for the pressure gauge and lights for the sight glass for running at night. Since I am not a scale modeler, I don't have a headlight on the tender (as that is where the engineer sits for the 2-4-0 and 4-6-0 and where a passenger can sit on the 2-8-2), or any lights on the running boards or for the ground and don't have any marker lamps on the tender. The headlights are not made from castings that are available and the classification lights and cab lights are not made to the full-size style.

The headlights are the top end of a swivel head flashlight mounted in a cast iron (the material which we had on hand) sleeve. This head uses an audio cable connection and for the 4-6-0, a four D-cell battery pack is kept in the tender. The classification lights

***Opposite, top:* Figure 156. Side view of the homemade classification lamps and headlight. *Opposite, bottom:* Figure 157. Front view of the homemade classification lamps and headlight.**

1908
1908

are made out of ¾" square tubing cut about 1" long. I welded a top on the tube and drilled three ⅜" diameter holes in three of the sides. Clear plastic was cut and glued to the inside of the tubing. A hole is drilled in the top and a 5–40 socket head cap screw holds the classification light to the base, which is made out of 1" aluminum angle. Each marker light has a 12-volt grain of wheat light bulb (like what was used in doll houses 12+ years ago) that is operated off a nine-volt battery in the cab. The classification lights are on the same circuit as the gauge lights.

The pressure gauge has a 12-volt grain of wheat bulb underneath the glass at the base of the dial. The cover was removed and I had to trim the steel casing a little to make room for the wires. The glass above the light was painted black so the engineer wouldn't have the light shining in his eyes. This does a good job of lighting up the gauge to see the pressure. It is not so bright to be able to read

Above: **Figure 158. The grain of wheat light bulb is under the glass of the pressure gauge.** ***Below, left:*** **Figure 159. Three grain of wheat light bulbs are mounted to the piece of aluminum angle behind the water glass.** ***Below, right:*** **Figure 160. The light bulbs light up the back side of the glass so the water level can be read when it is dark.**

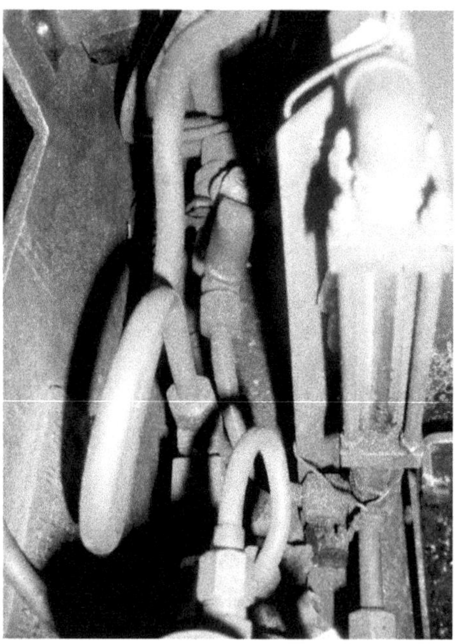

Figure 161. The grain of wheat light bulbs (covered in dust/ash) are seen in the angle behind the 2-4-0's glass.

if it is 120 or 121 psi, but does a good job of illuminating the gauge.

Behind the glass is a piece of alunimum angle (¾" angle on one engine, 1" angle on another) and holes are drilled on each side of the apex for the wires. One wire from each bulb goes through each hole. The holes are located to place the three bulbs evenly along the length of the glass. The aluminum angle is wired to the glass assembly to keep it from moving around. The lights do a good job of lighting up the glass. The water level in figure 160 shows the water level is just above half a glass. The water gauge, pressure gauge and classification lights are powered off of a nine-volt battery (which has a switch) in the cab.

Dynamo

Around 2000 I started messing around with making a steam powered dynamo for the 2-8-2 to power the lights. The dynamo that came out of this isn't perfect but can generate 6.2 volts at 1.1 amps.

Originally it was used to power the headlight, pressure gauge and water gauge lights.

Left: **Figure 162. This is the back side of the dynamo. The electric end is on the left and the steam end is on the right. The exhaust for the steam is mounted on the back side.** ***Right:*** **Figure 163. This is the front side of the dynamo. The steam side is on the left (steam supply enters in the front, bottom) and the electric side is on the right.**

When we changed over to the 12-volt grain of wheat bulbs, the 6.2 volts from the dynamo wasn't enough so the grain of wheat light bulbs (gauges and classification lights) are powered by a nine-volt battery.

The steam end is made from a piece of 1½" diameter brass and the generator end is made out of a piece of steel turned to 1½". The shaft from the fan is connected to the shaft for the generator by a piece of ½" hex brass (what we had readily available) and is secured to the shafts by a 5–40 socket head cap screw.

The socket head cap screw in the generator end keeps the generator/motor from turning. The base is a ½" × 1½" piece of steel. Small flats were milled on the round stock and they are bolted on to the base from the bottom.

Figure 164. The steam fan is made out of strips of sheet brass soldered into slots I cut on a brass hex. A set screw secures the fan to the shaft.

The steam end was bored out to around 1³⁄₁₆" and the fan is made from a ½" hex and ¹⁄₃₂" brass stock. I cut grooves in the flats on the hex for the brass blades, cut the blades from the flat stock and soft soldered the blades to the hex.

The fan (which is around 1¹⁄₁₆" in diameter) is secured to the ⅛" axle by a 5–40 set screw. On either end of the fan I put washers to keep the fan from rubbing into the cap or body. To hold the cap on the end, I drilled and soldered two studs (around a 3–48 size) to the end of the body. The exhaust outlet was silver soldered to the body and was drilled and tapped for a ¼–40 MTP. The steam inlet is a ³⁄₁₆" copper tube angling up at the 5 o'clock position.

Note: I did not balance the fan, which causes the lights to flicker. If I were to redo it, I would cut the blades too long and then turn the blades in the lathe to get them to the right diameter and help balance the fan.

Figure 165. The steam cap is made out of brass and is held on by two studs.

The steam end cap is made out of brass and has two holes drilled for the studs (the nuts are to the right of the cap) and a hole drilled in the center for the axle. If I were to redo this, I would see about installing a bronze bushing in the cap for the shaft and would also use something to seal the cap on the body. The end cap does leak as I did not seal it with a gasket.

I used a 12-volt DC motor (bought from Radio Shack) for the generator. I bored out the piece of steel for a slip fit on the motor and bored the opposite end to allow the collar of the motor to stick out of the main bore. I milled a slot in the bottom for the wires and drilled and tapped two 5–40 holes to secure the cap to it. The cap is made from 1½" diameter steel and before the cap is installed, I put a piece of electrical tape over the leads so they can't short out on the end cap.

The generator works decently. Here is a video of the dynamo working: https://www.youtube.com/watch?v=AQLJzQVHiUU&feature=youtu.be. Since I did not fine tune it and get the fan balanced, the headlight does flicker. I do use it on occasion, but since it makes some noise, it takes away from the stalk talk.

Figure 166. The 12-volt DC motor is placed inside a housing and a slot was cut for the wires to exit the bottom of the housing.

***Left:* Figure 167. View from the steam end with the cap installed. *Right:* Figure 168. View of the electric end with the cap installed.**

Grates and Ashpan

When burning coal or wood, the fuel needs a surface (grates) on which to sit but not fall into the ashpan while allowing air to come through for combustion. On full-size engines, the grates were typically made of cast iron and would rock back and forth to clean and dump the fire.

Figure 169. This firebox has a grate area of about 45 sq. ft. and has finger grates. The front has been cleaned and fire is burning on the front, clean grates and the rear grates have been cleaned.

The finger grates in figure 169 have staggered fingers extending from the center of the grate and there is air space between the fingers. This shows the rear half of the grates after they have been cleaned from the previous day's ash. The front half has already been cleaned and the fire is begun on these grates. When looking at these grates, there is a lot more solid metal than space for air to enter the fire. From my experience with 1:8 scale coal burners, we need proportionately a lot more area for air to enter the fire than the full-size.

When we first built the 0-4-0 (later modified into a 2-4-0) we made grates (grate area is 48 sq. in) out of ⅛" × ½" stainless steel bars spaced about ⅛" apart. While she did steam decently most of the time, when she was under a heavy load on a long grade, the steaming qualities were less than ideal. While hooking up the engine and adding an arch did help, we decided to try and make new grates for the engine using $^{3}/_{16}$" × ½" stainless steel spaced 5⁄16" apart. This allowed more air into the fire and helped her steaming qualities. (We also added a stainless steel arch inside the firebox which further helped the steaming qualities.) With the larger spacing, the ash and impurities could drop through the grates more easily, which helped to keep the fire cleaner.

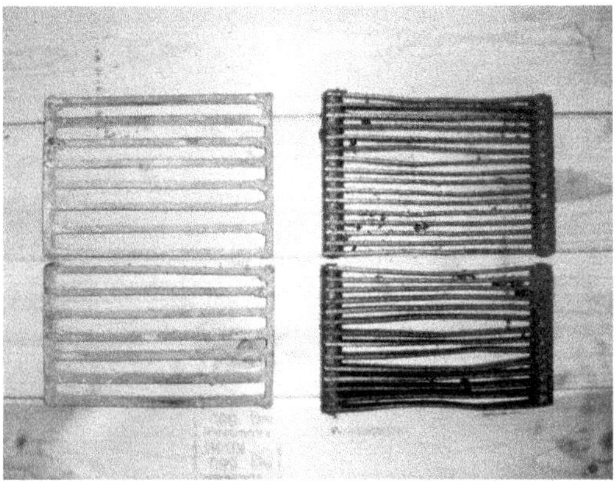

Left: **Figure 170. The old grates, with smaller spaces for air, are on the right and the new grates, with larger air spaces and bars, are on the left.** ***Right:*** **Figure 171. Bottom view of the 2-4-0's grates in service. Her grates install from under the mud ring.**

Figure 171 shows the 2-4-0's grates in use and unlike the 2-8-2 and 4-6-0, these grates are installed from the underside of the firebox. The original ashpan has been modified by allowing the ash to fall into the hopper (originally it was there for cosmetics) below. This increased the ashpan capacity and we did not have to worry about the ash building up as fast and touching the grates.

Figure 172 shows the grates (grate area is 150 sq. in.) in the 2-8-2. These grates are made from ¼" × ½" stainless steel with 5⁄16" space between the bars. The grates are made in three sections so they can easily fit through the firebox door.

In figure 173 the grate bearers and the top part of the ashpan (with ash still in it) can be seen, and in the center is the hopper.

Figure 174 is of the right side of the grate bearer and top part of the ashpan. When the ashpan needs to be removed, the hopper can drop out the center of the frame and fit through the opening in the

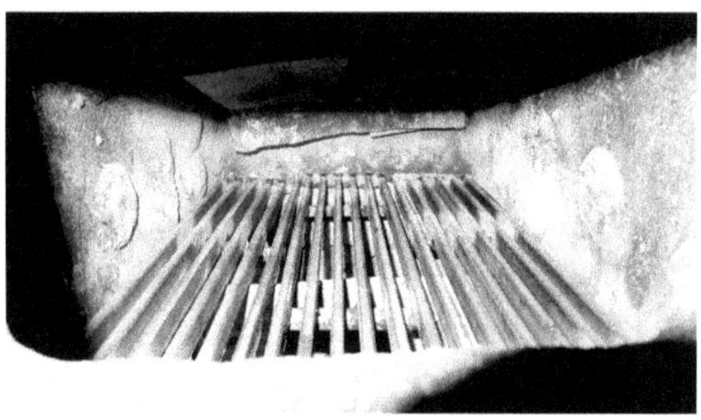

Right, top: **Figure 172. The 2-8-2 has a grate area of 150 sq. in. and her grates are in three sections.** ***Right, bottom:*** **Figure 173. The grate bearers are attached to the top part of the ashpan and the hopper is in the center extending down between the frame rails.**

Figure 174. The grate bearers can slide to the side and come out from under the firebox, after removing its retaining bolts and the injector.

trailing truck frame and the top part of the ashpan can be removed by taking off the right side injector, removing two bolts that hold the top pan/grate bearer in place (located under the backhead) and then it can be slid to the right side. The boiler was installed before the pan was installed so we could design the ashpan and grate bearers to be removed without taking the boiler off of the frame. When we tried to convert her over to oil, the oil pan slid in the same fashion; as a side note, the open ¼" tube is the steam line that went to the atomizers.

Figure 175 is of the grates (the grate area is 44 sq. in.) in the 4-6-0. These grates were made by LocoParts and are designed in two pieces to fit through the firebox door.

Figure 176 shows what is visible of the ashpan by the right rear driver. With the long narrow firebox, the pan is kept close to the mud ring to help prevent air from just

***Left:* Figure 175. The 4-6-0 has a 44 sq. in. grate area and her grates are in two sections. *Right:* Figure 176. Sheet metal was added to the sides of the ashpan to help seal it off and have the air enter through the back of the ashpan.**

drawing into the front of the firebox and not through the back. Since the metal frame does not fit right against the mud ring, I added strips of sheet metal to the metal frame to create a better seal along the mud ring.

Figure 177 shows the front of the ashpan and the bolts holding it in place. It is designed in two parts: the top is the grate bearer and the main frame, and the bottom is the hopper. Note: on the back of this wheel the line between the tire and wheel center is easily seen.

Figure 178 is the rear view of the ashpan with the bolts and bent ⅛" thick bar stock holding it in place. When the drawbar pocket and the bars holding up the ashpan have been removed, the ashpan can drop far enough to remove the grate bearers by pulling them to the rear of the engine, and then the hopper, which straddles the rear driving axle, can be removed. When we want to change the engine over to burn propane, the burners (also from LocoParts) are installed in a similar fashion. The change-over usually takes about 45 minutes to an hour to complete.

***Left:* Figure 177. The front of the ashpan also had sheet metal applied to it to help seal off the ashpan. *Right:* Figure 178. The ashpan has a sliding door (it pushes forward to line up with the openings in the door for the front and rear hopper) and is pinned in place by a long drawbar pin. The air enters into the ashpan through a 1"+ gap between the mud ring at the door sheet and the ashpan.**

The light seen from the firebox is the light bulb that I place in the firebox to keep the boiler warm and keep condensation from forming when she is in storage. The back is open to allow air travel across the entire grate (from the rear to the front) and not have air enter in the front and bypass the back of the fire.

With these grates and ashpan set-ups, the engines all steam very well and any part of the coal burning arrangement can be removed without needing to take the boiler off of the frame. Since the 2-4-0 is the smallest (the boiler only holds about two gallons at running level) she is also the least forgiving. While pulling a four- or five-car train at the White Creek Railroad, she would steam pretty well until a couple of the long grades with the old style of grates. With the new style of grates (this was the last in the line of improving the valve gear, drafting and arch) she did a much better job keeping up pressure going up some of the long grades. With the new grates, it was easier to inject water going up the grades and not have the pressure drop more than 3–5 psi and have it rebound back again. Since the 2-4-0 is a small engine, adding water and adding coal worked best when they were alternated and only done in small quantities but at more frequent intervals. With only adding a small amount of coal or water, the firebox temperature and the water temperature in the boiler were allowed to remain relatively consistent, which meant that the rebound times were minimal.

With the 4-6-0 and 2-8-2 being quite a bit larger than the 2-4-0, they are also more forgiving if extra water or coal is added at once. Even with the larger engines I prefer to add water and coal in more frequent, but smaller, amounts. In the larger engines, it may be possible to see into the firebox to look at the condition of the fire. If the coal has long flames, it still has life left. If the flames are getting short or the coals are just glowing red, they are on the tail end of their life. Those areas that are on their way out are the areas that should get fresh coal first. In most miniature engines that I have seen, coal is shoveled into the firebox door and then pushed around with a poker stick to keep the fire level and spread the fresh coal around.

Each of our engines fires differently. With the 2-4-0 and 2-8-2 I can add coal and push it around with the poker stick and the fire will generally be good for running around 1,000–1,500 feet. Then I use the poker stick and break up any coal that has fused together and I might have to add more coal. The 2-4-0 likes a thicker fire than the 2-8-2. The 4-6-0 also likes a thicker fire, but the long narrow firebox likes to have the fire really raked and dug into to break up any coal that has fused together, especially in the back half of the firebox with breaking and digging down to the grate. If I do more gentle raking like I do on the 2-4-0 and 2-8-2, the steaming capacity is greatly diminished and though she is a fair steamer she will drop from 130 psi down to 115 psi climbing a grade with a decent load. With really raking the fire every 1,000–1,500 feet she is a great steamer and might drop down to 125 psi but the pressure will start to climb and sometimes pop the safety while still climbing the grade when pulling her maximum load.

Figure 179. This 15" gauge 4-4-0 was originally built in the 1920s by the Cagney brothers of New Jersey. She supposedly ran on Coney Island and now runs at the Strasburg Rail Road.

When teaching people how to run the 15" 4-4-0 at Strasburg (which burns anthracite coal) I let them try adding a lot of water and a lot of coal at once to let them see what happens. They usually start out well, but as the day goes on, the lag time from rebuilding pressure (either from adding coal or adding water) catches up to them and they start getting behind. Before it gets too far down that slippery slope I will go over adding a little bit of coal and water at a time (but more frequently) and explain that it helps keep the temperatures more constant and in the end, adding a little at any one time will be less of a headache. It typically doesn't take them too long to get the hang of it and at the same time they can recognize the symptoms of adding too much coal or water in one shot and know how to correct it easily. Firing an engine (miniature or full-size) is an art form that depends on the condition of the fire, water level, pressure and the demand of steam that will be needed.

Learning Oil Firing on a 15" Gauge 4-4-2

When Dad started in the live steam hobby in late 1989, he bought an oil-burning Shay locomotive that was built by Dutch Boysen of Yakima, Washington. This engine burned oil in a pot burner where the oil would be vaporized by the hot metal of the pot burner and then it would ignite. The three engines (2-4-0 modified from a RRSC 0-4-0, RRSC 2-8-2 and Allen Models 4-6-0) that Dad and I built were built as coal burners. Around 1998 we tried to install a pot burner on the 0-4-0. While we could make steam, we were not able to make enough steam to run much so we changed her back to coal. Around 2000 we tried the 2-8-2 on oil (so we could run when everything was really dry) and made twin atomizers and an oil pan based off of the RRSC plans. The oil pan and burners could be inserted by sliding them in from the right (engineer's) side of the firebox. The atomizers did work, but one of them liked to go out (most likely due to the oil hole not being perfectly centered in the atomizer) and because we did not care for the smell, we changed her back to coal.

In the summer of 2004, Dad and I started working on the Spirit of Traverse City, a 15" gauge 4-4-2 that ran around Clinch Park Zoo (see figure 84; after the zoo was closed it ran around Clinch Park) from 1982 until 2011. She was an oil burner and had a flat atomizer where oil would run out over a ledge and be atomized by a jet of steam underneath it. The burner worked, but it was not the best as not all the fuel was atomized and there was not enough air entering the fire so the flame was always a dull reddish to orangish color and it was difficult to keep pressure up when hauling a train that was only half full.

When the boiler was retubed in June 2005, we decided to take another stab at making an atomizer and oil pan and try to improve the steaming qualities of the Spirit of Traverse City. For the new atomizers I modified Chet Peterson's atomizers for 1:8 scale and enlarged them without making them twice as long (see fig. 86). (The drawings are in the "Large Scale Atomizers" section earlier in this chapter.) They worked great when we tested them on air which was a nice shot in the arm.

The old oil pan was made out of ⅛" thick steel and was filled with firebricks. The only air inlet was a small area around the burners (plus a secondary air supply from the door) and the pan was bolted to the mud ring and seemed like it was assembled in sections. We wanted an easier way to get the oil pan on and off the engine and we noticed that if the rear cab support was removed, there was a straight, clear shot (the firebox was supported on the front and not on the rear) to slide the oil pan under the firebox from underneath the cab. We had angle iron spot welded to the mud ring so there was ⅜" space between the mud ring and the angle to allow the new oil pan (made from ¼" thick steel) to slide in easily. We talked with several people who have operated oil burners (miniature and full-size) and the basics of making the oil pan were: (1) make it as large (volume) as practical; (2) have the main air supply (and damper) come from underneath the fire in the rear of the pan; (3) have a secondary air source in the firebox door; and (4) if practical, add holes to the sides of the pan in the rear half for secondary air supply (this one was not practical for us). Later we enlarged the main air supply in the rear of the pan so that the area for the air supply was close to the net area of the tubes and also added the damper once we knew how we could hook it up easily.

Figure 180. The blower valve is handle 1, the atomizer steam valve is handle 2 and the large handle 3 is for the firing/oil valve.

The cab of the Spirit of Traverse City had two different color valve handles for the blower (red) and the atomizer steam supply valve (green) located on the fireman's side of the cab. They had a mark (a ring on one and a blot on the other) to show how far open each valve was. A large black handle serves for the oil valve (a ball valve located under the cab) and has a quadrant with notches so a fine adjustment on the valve can be noticed. There was no way to pin the handle to the quadrant, but we made marks on the quadrant for a spot fire, easy running and pulling the train up the 2.5–3 percent grade. (The black handle on the quadrant is the fuel control valve. All the firing controls were situated together.)

The air valves (green for the atomizer, red for the blower) and the connection for an air line are located below the fireman's side of the cab. Their handles are also painted green and red and they are used when the engine is being fired up. Below the cab a glass bulb filter was added to the fuel line to collect debris so it would not clog up the atomizer.

Down the line from the filter is the oil control valve (a ball valve) and the rod from the valve handle is connected to the firing handle in the cab.

The firebox door was modeled after the prototypical firebox door for an oil burner. It is made from mild steel and has a sight hole covered by a round cover (which can be opened easily) on the rear of the door and on the bottom it has a rotating damper to allow a secondary source of air to enter the fire through the firebox door. She fired the best when the damper door was open at a 45-degree angle.

Top: **Figure 181. Underneath the fireman's side of the cab is the air hook-up for the blower and atomizer. The fuel line, glass bulb filter and valve are also under the fireman's side of the cab. In front of the air hook-up is a $\frac{3}{8}$ NPT Penberthy injector. We lowered it below the running board (and clear of the atomizer lines) to be used as a non-lifting injector.** ***Bottom:*** **Figure 182. The firebox door has a sight hole (covered by the round cover that can be swung to the side) and a damper for the secondary air supply and is modeled after the full-size doors for oil burners.**

In addition to improvements to the firebox end, we also decided to improve the front end. The original blast nozzle was too large and part of it was blocked off to make it smaller and create a stronger draft by increasing the velocity of the exhaust. Since it was unknown what size the blast nozzle should be, one was fabricated using a ¾ NPT coupling (wrapped a ¼" tube around the coupling and drilled holes in it for the blower) and a drilled plug for the blast nozzle. We drilled out several plugs to different sizes and after we would try one we would turn the fire off, open up the smokebox and change out the plug to another size and relight the fire. The new blast nozzle and raised petticoat are seen in figure 91.

Over the first year of firing with this new arrangement, several adjustments had to be made. The burner was aimed up towards the bottom of the firebox door as in the original lower position, some of the spray was hitting the bottom of the pan. Figure 88 shows the burners in action and angled up towards the door. The air supply in the rear section of the pan was increased and a damper door was added to the main air supply and also to the air that came around the burner. (Later we learned that the air supply around the burner should be minimized, as when the locomotive is drafting, the air will be pulled in around the burner and directly into the flues. This air won't really help combustion and won't be heated and there is a good chance that over time this can thermal shock the flues and they will start to loosen up and leak.) We also had to adjust the petticoat so it gave the largest distance between it and the blast nozzle as possible. When it dropped, the draft would be less and the engine would not steam as well.

Figure 183. Handles (painted with grey primer in this picture) were used to control the dampers. The top handle controlled the front (by the burners) damper and the bottom one controlled the rear damper.

To control the draft, dampers were added that could let more or less air into the fire. One set of dampers was in the front controlling the air supply by the burners. The second one was in the rear of the floor of the pan. The front dampers were controlled by the upper handle (in grey primer paint) and were kept about one-third of the way open. The bottom handle controlled the rear damper

A homemade universal joint was made to connect the handle to the angled all thread connection leading to a crank just above the trailing truck wheel. These are all painted in a grey primer. We used two pieces of ⅜–16 all thread and a long nut. The all thread was rigidly attached to the crank end and to the handle/universal joint end. The long nut was only rigidly attached to one of the pieces. When the handle was turned, the long nut would thread on or off the other piece of all thread, which would open or close the damper. This damper was typically kept wide open, unless a spot fire was needed during long waits while waiting for passengers to show up. (Note: The straight all thread with the homemade T handle connects to the blow down valve.)

Firing up the locomotive using compressed air was quite easy. We used a piece of stove insulation

Above: **Figure 184. The handles are rigidly connected to the cab floor support and homemade universal joints connect the controls to the dampers.** ***Below:*** **Figure 185. The angled threaded rod connects to the arm to control the front damper. The rear, long-threaded rod spins with the handle and threads the long nut on and off the short all thread that connects to the lever to open or close the damper. The level all thread connects the blow down valve (a ball valve) to the T handle on the back side of the cab wall support.**

on the end of a ¼" rod soaked in diesel oil to light the fire. We would put the burning insulation by the burners and open the air supply, but not enough to blow it out. Then we would open the oil valve and feed some oil to the atomizer and typically the burner would light right away. During the firing-up process, the burner was kept low to keep the firebox and boiler from being thermal shocked and give a good even heating up. A good rule of thumb is for every foot diameter the boiler is, one hour of fire up is needed to go from cold to full boiler pressure. Changing from air pressure to steam was a bit tricky since the lines are cold and the steam will condense and blow water out the atomizer, sometimes putting the fire out. If we cracked the steam valve to the atomizer while the air valve was still open, the condensation would generally blow out from the atomizer without putting out the fire. Once the condensation had blown out, we would open up the steam atomizer valve while shutting off the air valve.

Once the atomizer and blower were warmed up and the boiler was up to operating pressure, we would finalize the setting of the atomizer and the blower. We would keep the blower turned up enough to provide a good draft through the firebox, but no higher, as the higher the blower, the more heat is being sucked out before it can be absorbed by the boiler. The steam supply to the atomizer was turned up high enough to produce a good fire. If too much steam is used, that extra volume of steam reduces the amount of air (and oxygen) that the fire can get.

When we started running the Spirit of Traverse City, we had to remember to bump

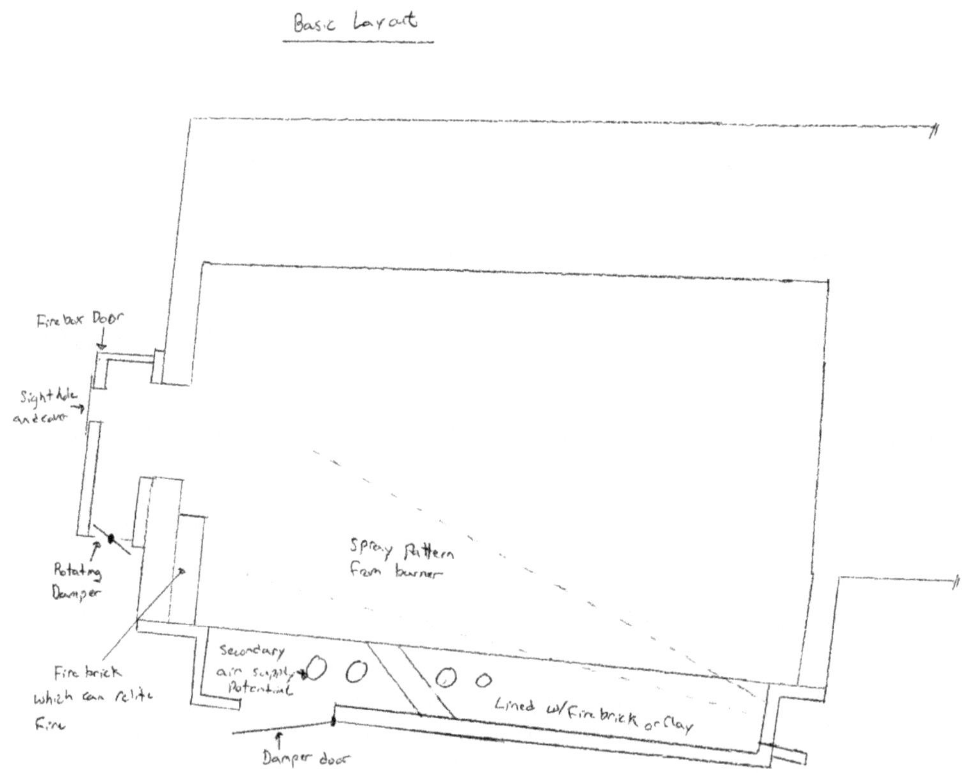

Drawing 40. This is the basic layout for the fire pan, air holes, dampers, burner angle, flash wall/brick at the door sheet and the door.

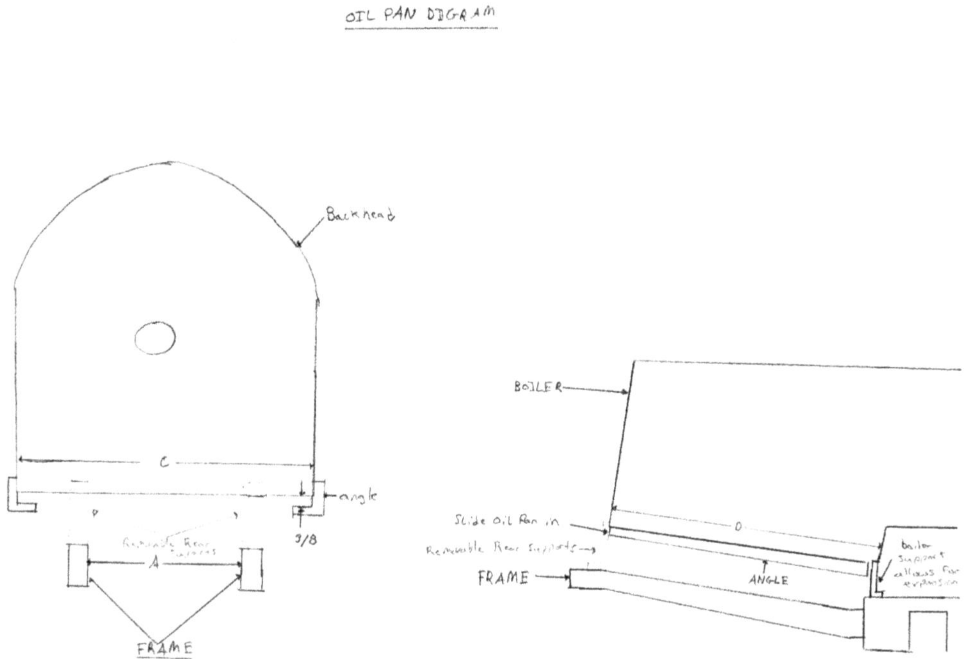

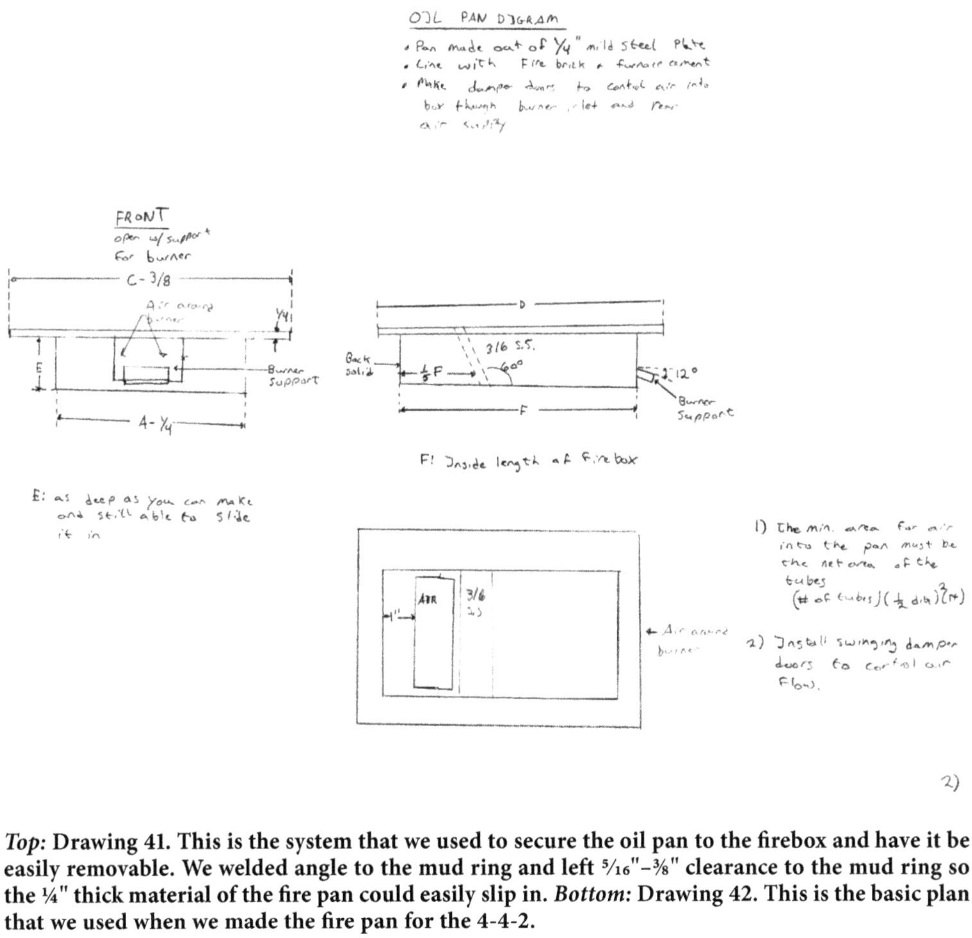

Top: Drawing 41. This is the system that we used to secure the oil pan to the firebox and have it be easily removable. We welded angle to the mud ring and left 5⁄16"–3⁄8" clearance to the mud ring so the ¼" thick material of the fire pan could easily slip in. _Bottom:_ Drawing 42. This is the basic plan that we used when we made the fire pan for the 4-4-2.

up the fuel supply to the fire before or as we were pulling out the throttle (or the fire would be chuffed out) and to bump the fuel supply down as the throttle was being closed. After losing the fire or having too much fire, which would come out the air supply vents, we got the hang of it. When running up the hill, we usually closed the blower as the exhaust created plenty of draft for the fire. When the water was added in smaller, more frequent amounts and the fire was kept so the stack had a slight haze, she could run five trips around the 0.4-mile track while pulling a fully loaded train and not drop below 140 psi (MAWP is 150 psi) coming up the 2.5+ percent grade.

While running with oil is nice in the regard that the engineer has more control over the fire and can back off the fire while sitting in the station (see fig. 87), I prefer to use coal in the 1:8 engines, but in an amusement ride setting I would rather have a good oil burner.

The diagrams in drawings 40–42 are what I found to work well with the oil pan and burner.

Propane

The main fuels I have seen used on model steam engines are coal, oil (diesel or kerosene) and propane. Each fuel has its advantages and disadvantages. Good coal can be hard to get and it produces ash that needs to be disposed of. Diesel oil is readily available and kerosene can be bought, but in fewer locations. Oil also produces soot and dark smoke and it has a distinctive smell. Propane burns cleanly and does not really produce much smell. The propane tanks can be filled at any place that deals with propane. For the most part, propane is carried in a separate car behind the tender.

As far as efficiency with propane versus coal or oil, I am not sure as I don't know any exact numbers. I believe propane has a lower temperature and BTU than oil and coal. Unlike coal where more draft equals more heat, with propane, less draft equals more heat.

With coal, the blast nozzle is 14–18 percent the diameter of the piston. With oil it is 18–22 percent the diameter of the piston (less draft) and I would think with propane it would be closer to 25 percent of the piston diameter. I have seen people with propane burners have the petticoat come down to the blast nozzle or have the blast nozzle picked up to the petticoat to minimize draft.

Around 2002 we converted our Allen 4-6-0 over to propane to be able to run in the dry seasons without starting up any fires. Our ashpan, oil pan on the 2-8-2 when we tried that and propane burners were all installed after the boiler was on the frame so that way we could change them out easily and have them designed to be installed and removed with the boiler on the frame.

I extended the petticoat down so it was ½" above the blast nozzle by using a strip of sheet metal, bending it around the existing petticoat and using a large hose clamp to clamp it around the existing petticoat.

We used burners that were made by LocoParts and made a plate out of sheet steel that would fit over the burners and limit the amount of extra air that could enter the firebox. Propane gets the air that it needs through the mixing process of the burner. Any air coming in around the burners is not needed for combustion, and this cool air will act to cool down and perhaps thermal shock the firebox.

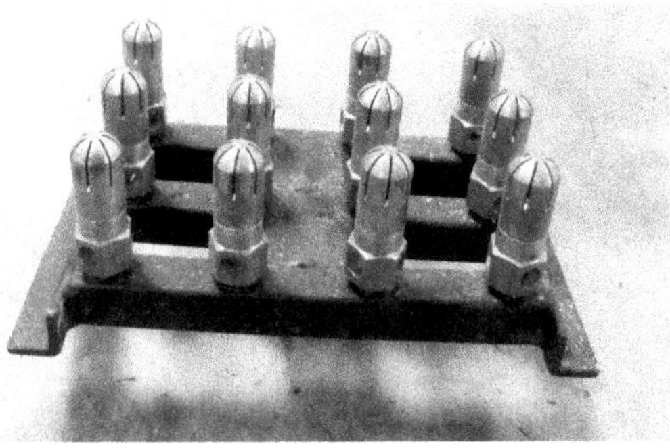

Left: **Figure 186. We bought the burners from LocoParts and the 12 burners arrived attached to the manifold, ready for us to install them on the engine.** ***Right:*** **Figure 187. The LocoParts burners arrived attached to the manifold.**

Figures 186 and 187 show the burners that we bought from LocoParts for the 2-4-0. Figures 188 and 189 show the sheet metal plate we installed (above the air inlet in the burners and below the flames) to minimize the amount of excessive cold air entering the firebox. We could easily slide this up into the firebox once the ashpan was removed.

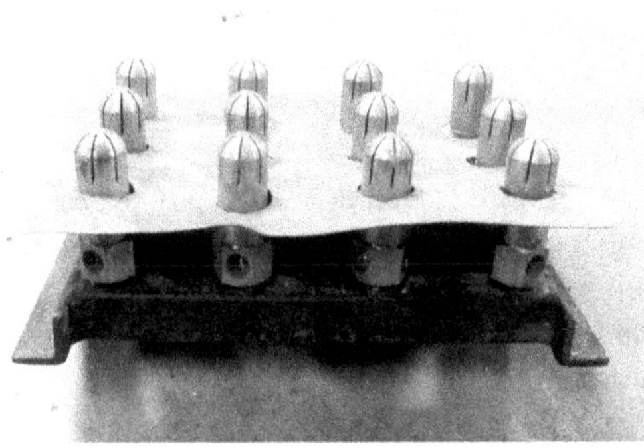

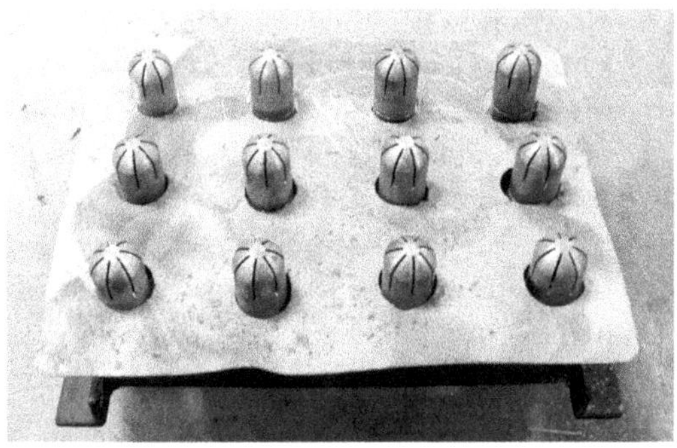

Left: **Figure 188. To help minimize the air flow into the firebox, we made a cover using a piece of sheet metal to set over the burners, above the air opening in the burners.** ***Right:*** **Figure 189. Another view of the sheet metal cover over the burners.**

We flattened a piece of stove pipe and drilled holes in the burner locations and placed it over the burners to not allow any cold air to enter into the firebox around the burners. This piece is kept below the slits for the flames so it does not interfere with the burners. We ran the 2-4-0 on propane a couple of times and then changed her back to coal. This was to see how easily we could do it if we had to do it later (for example, if we couldn't find good coal or had a long dry spell). We didn't tweak the set-up with her at that time.

We had a control valve and pressure gauge on the engine so we could control and see how much psi we needed with the burners. The propane was carried on an open bottom flat car (propane is heavier than air, so this will allow any propane that leaks out to be free and not collect). Dad bought two horizontal propane tanks and had the propane shop refit them so they would draw propane gas instead of liquid propane.

The flexible lines between the cars/tender and engine should be rated for propane and the line under the car/tender should steer clear of any potential pinch points around the trucks. If the tender or car derails, the propane line should be free of any danger of being pinched or punctured by a wayward truck. I would also highly, highly recommend (if it is not already required by clubs) to have safety chains between all pieces of equipment that have the propane line. The chain and connectors should be strong enough to take a high force spike if or when a coupling fails and the entire train weight pulls on the chain and connections. Even if drawbars are used, I would still recommend safety chains as a backup. Drawbars and their pins can also fail.

The modifications worked well. We ran her on the White Creek Railroad with a decent load up some 2 percent+ grades. For propane, she needed around 3–5 psi with light work, 7–10 psi on medium grades and around 12+ psi on steep grades. She did not steam as easily as she did with coal, but then again we also did not tinker with propane all that much and could have gotten better results if we continued to mess with it.

For using propane on a large engine, like a Mikado, I would recommend contacting LocoParts. They will need to know the grate area (if it's a RRSC 2-8-2, the firebox is 10" × 15" and has around 22 × ¾" ID flues and carries 8 to 8½ gallons of water at running level).

When Dad was thinking about trying to easily convert the 2-8-2 to propane, he didn't see how a single manifold/burner set-up (which is the size of the grate area) would fit in there without removing the boiler from the frame. However, if three manifolds could be made, they should be able to fit up through the frame center and then be positioned and held in place by some kind of frame.

I would recommend trying to design the firebox end so things can be changed out easily. Our ashpan can drop out easily by removing three bolts from the side frame around the trailing truck. The frame for the grate holders and sides of the ashpan need to have two or three bolts removed from the back (under the cab floor) and then it can be slid out the engineer's side once the injector has been removed. The oil pan was made to slide in this same way.

Propane, like oil, can be directly and instantly controlled from the cab. This means that the engineer/fireman can instantly change how much fire there is and how intense the fire is very quickly. This is not always a good thing as the firebox sheets can be thermal shocked (which over time can cause cracks in the sheets and seams and leaks in the sheets, around the staybolts and around the flues) from drastic changes to the intensity of the fire. Just because the fire can be turned up when an injector is turned on to keep the pressure from dropping doesn't mean that it should be. The pressure should behave in a similar fashion to a solid fuel or coal fire. With coal, there is a lag time between adding coal and turning up the blower to getting the extra heat. This should be mimicked with propane and oil. For firing up, the fire should be kept low to allow the boiler and firebox to heat up slowly and evenly. A general rule of thumb is the diameter of the boiler in feet is the fire-up time from cold in hours. So a 6" boiler should have a 30-minute fire-up time and a 8" boiler should have about a 45-minute fire-up time. Now an 8" boiler with two gallons of water will fire up faster than an 8" boiler with five gallons of water. The point is to bring the boiler up to pressure slowly and evenly. An 8" diameter boiler holding two gallons should *not* be fired up to full pressure in 10–15 minutes. A much more reasonable fire-up time for that size boiler is 25–30 minutes.

Long Track, or a Short Loop

Once, I was asked if I prefer running in a short loop or a large track like the White Creek Railroad (located just north of Grand Rapids, Michigan. This private track has over five miles of main line and over six miles of track total). Somehow I get the feeling that my answer took them by surprise when I said, "short loop." I then explained that if I want to run, all I need to do is get my engines out and run on my short 485' loop that I have going around my house. To run a longer loop at a club means packing up the equipment, at least a two-hour drive one way, unload, fire up, run, shut down, reload the equipment and drive home again. I enjoy running more when it is not an all-day project and I can do it easily so I will run more frequently.

Growing up in Traverse City, Michigan, Dad and I would take trips down to the White Creek Railroad. It was about a drive of about two hours and 15 minutes and the whole day would end up being about 12 hours. It was fun running on that long of a track, but we didn't do it frequently due to the drive; we visited the White Creek maybe three times a year. Now that I have my own track, I am able to run and put the engine away in a matter of three hours and typically run the train once every other week during the summer.

If there were a large track close to me, it might be a different story. I enjoy running on a large track and being able to take different routes and pull trains of different lengths and heading up S-curves. I'm not sure if I would want to have a large track on my property (if I could, that is) as 485' of track gives me more than enough track work. I really don't have time for even more track work!

The Bigger the Better, Right?

When Dad got started in the hobby he was looking for an engine that would take tight curves as a loop that could fit in the backyard would need to have 20'–25' radius curves. Because there was a 7.5" gauge track nearby (Bill Fitt's) Dad got started in 1:8 scale. In 1989 Dad bought a small two-cylinder Shay and started working on a Railroad Supply 0-4-0. The Shay was small and didn't run very well, which we thought was due to our lack of experience. As we learned more, the we discovered that in truth there were a good number of problems with the Shay. Once the 0-4-0 was running, the Shay was put on the back burner.

We put many miles on the 0-4-0 from 1996 through early 1999. (I made the lead truck for her around 2000 or 2001.) In late 1996 Dad bought castings for a Railroad Supply 2-8-2 and in May 1999 we lit the first fire in her. She was the main engine that we ran once she was built, but due to her size (900+ pounds) it was a pain to get her back on the track when she derailed. She is a great engine, pulled some long trains at the White Creek Railroad (the longest was 31 cars) and we had a great time with her, but I was nearing the end of high school and Dad wanted a smaller engine that he could run by himself while I was at college.

In September 2000 we started building an Allen Models 4-6-0 and she was finished in May 2001. She is a nice mid-size engine that one person can move around and rerail when the engine derails. Growing up, we ran the 0-4-0/2-4-0 and 2-8-2 a lot more than the 4-6-0 as we only had a little more than one year after the 4-6-0 was finished before I went off to college.

I have all three engines at my track here in Pennsylvania. While the 2-8-2 is impressive, fun to run and a good engine, it is really too big for me to get out and run by myself (my kids are not old enough or strong enough to help yet) and is only run when my parents are here or I have some live steam friends over to help.

The 4-6-0 is also a good running engine and if I have a few friends coming over, I will pull her out and pull all five cars that I have. She does not pull as much as the 2-8-2, but still puts on a good show. I also will get her out when the kids want to ride on the train.

I find myself getting the 2-4-0 out the most as it is the easiest engine to get out by myself. When pulling the family, I don't need one of the larger engines and the faster fire-up and shut-down times means less time is spent getting it ready and putting it away. Since the 2-4-0 is small, she is less forgiving than the larger engines and somewhat more challenging when pulling a decent load up the 3 percent grade, but I also enjoy that aspect as well.

Figure 190. The 2-4-0 is run the most on my track. She is the smallest engine and the easiest to get in and out, quickest to fire up and shut down and can pull a 1,500+ pounds train up the 3 percent grade on a 52' radius curve. The train can consist of engine, tender, engineer, two gondolas, two flats and three adults or six kids, which is plenty for what I need. Veronica is taking our boys for a ride in this picture from the summer of 2015.

I know several people in the hobby who will run their diesel more than a steamer as it is easier, or who will run their smaller engines (steam or diesel) as they are easier to move around and get back on the track if they derail.

Equalized Brakes

Note: The information in this section can be adapted to any steam locomotive of any gauge.

Full-size engines use a series of equalizers/levers in the brake system (similar design to the equalizing for the spring rigging) so each of the brake shoes receives the same amount of force. Some scale models do this, some don't. With a steam engine, since the driving wheels are connected by solid rods (which transmit the power from the cylinders to all the drivers) it would be possible to have only one driving axle have the brakes and have all the braking force applied to one driver and the braking force would be transmitted to the other drivers through the rods. The brake shoes would wear quite quickly as they will be taking the abuse for all the drivers and the wheels with the brake shoes will most likely wear faster than the others.

Another way would be to have brake shoes on all the drivers but they are connected directly to each other through solid tie rods. The force from the brake cylinder pulls on the entire solidly spaced brake rigging and applies the brakes. Each wheel more than likely won't have the same amount of force as if the tie rods are not the same length, more force will be applied with shorter tie rods than longer tie rods. As the shoes wear, they will most likely not wear evenly which will put more force on the shoes that wear less. If one brake shoe is replaced, the tie rods will need to be readjusted so that when the new brake shoe is bearing against the wheel, the old worn brake shoes are also bearing against the wheels.

With an equalized brake system, the force from the brake piston is transmitted through a series of levers which will direct even force to all the brake shoes. As one brake

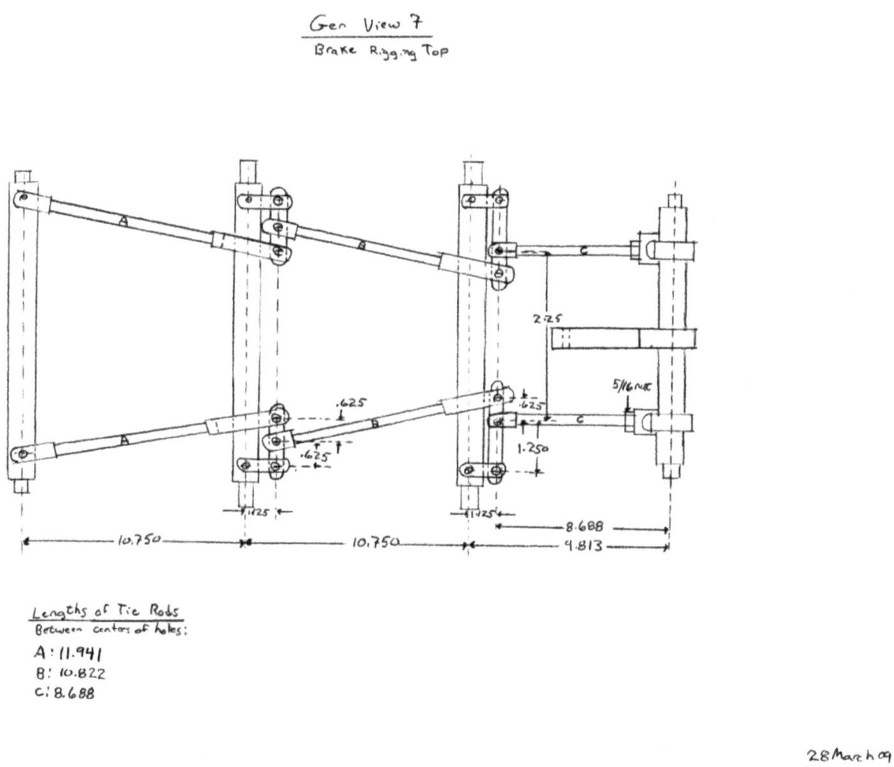

Drawing 43. With equalized brakes, the pull rods don't attach directly to the brake beams but to levers/equalizers that evenly divide the force between all the wheels.

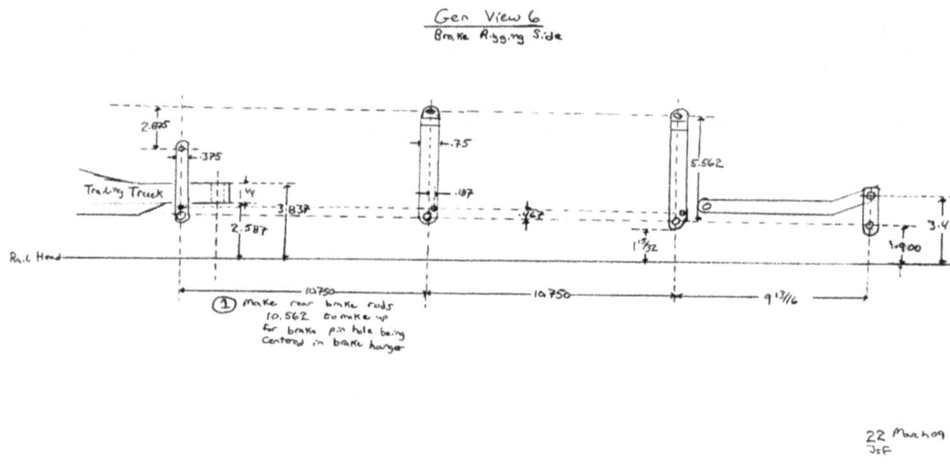

Drawing 44. When designing an engine, the brakes also need to have ample clearance above the rails to minimize damage when the engine derails.

shoe wears more than the others, the levers are free to float and change their positions to allow all the brake shoes to contact the wheels and have the same amount of force.

Drawings 43 and 44 show an equalized brake system for a 4-6-4 that is being designed and built. The brake cylinder is at the front of the engine and the lead brakes are the first ones to receive force from the cylinder. The first equalizer is a 1:2 ratio so one-third of the force is directed to the front brake shoe and the remaining two-thirds are sent to the rear brakes. The second equalizer is a 1:1 ratio so half the remaining power (one-third of the full power) is sent to the second brake shoe and the other remaining half (one-third of the full power) is sent to the third brake shoe. Also, when designing

Left: **Figure 191. At the bell crank (connected to the brake piston and changing the vertical force of the piston to the horizontal pull on the brake rods) is a clevis that has a 1:1 equalizer. From that equalizer the tie rods split for the left and right sides. As one side wears, the first 1:1 equalizer is allowed to swivel and correct for the uneven wear in the brake shoes.** ***Right:*** **Figure 192. The first tie rods head back to the first equalizer. This has a 1:3 ratio so three-quarters of the force is transmitted to the remaining drivers and one-quarter of the force is transmitted to the first brake beam and brake shoe.**

the brake system/rigging, extra clearance should be built in above the railhead to minimize the damage to the brake rigging when the engine derails.

The drawback to the way I have this drawn up is that the tie rods that connect the bell crank to the first equalizer are solid and are not allowed to float when one side wears more than the other.

A Railroad Supply 2-8-2 uses an equalized brake system that fixes that problem. Figures 191–195 show the equalized brakes on the 2-8-2.

***Above, left:* Figure 193. The next tie rod heads back to the next equalizer, which has a 1:2 ratio. Two-thirds of the remaining force (half of the full force) is transmitted to the next equalizer. One-third of the remaining force (one-quarter of the full force) is transmitted to the second brake beam and brake shoe. *Above, right:* Figure 194. The next tie rod heads back to the next equalizer which has a 1:1 ratio. Half the remaining force (one-quarter the full force) is transmitted to the last brake beam and shoe and half the remaining force (one-quarter the full force) is transmitted to the third brake beam and brake shoe. *Right:* Figure 195. The last tie rod connects directly to the fourth brake beam, which receives one-quarter the full force of the braking power.**

The amount of force needed from the brake rigging depends on the type of materials that are used. The most common materials that are used are cast iron (wheels, brake shoes), steel (tires/wheels) and cast brass/bronze (brake shoes). Most kits already have the braking power needed worked out and are designed that way. If designing or modifying an existing design, it will probably take a bit of trial and error to get the maximum force needed. When the brakes are fully applied, the drivers should not slide under normal (dry) conditions.

Tender Brakes on the 2-4-0

Note: The information in this section can be adapted to any steam, diesel, electric or gas locomotive of any gauge.

In the early 2000s Dad and I looked into putting brakes on the tenders to increase the braking capacity. When pulling long trains at the White Creek Railroad, we would choose our routes so we didn't have a long, straight downhill; we wanted curves to add resistance and help keep the train from going too fast. Due to the Mikado being heavier, it had better braking capabilities and we also tended to run her more as the engine weight (and braking power) was not as small in proportion to the train weight as compared to the 2-4-0 or 4-6-0. In the early 2000s we were eyeing air brakes but as these needed a small air compressor, tank, battery and car for it along with valves, and the truck not having much room for an air cylinder under it (connected to both brake beams of the truck) we decided against it. We also didn't have a good understanding of how the foundation brake rigging was built on the full-size so we didn't consider that either.

When I decided to add brakes to the tender, I wanted something simple to use. Some people I know have hand brakes or vacuum brakes on the tender and would need to operate them separately from the engine brakes. I wanted the tender brakes to work in tandem with the engine brakes (which are steam brakes) so I decided to have the brakes on the tender be powered by a single cylinder. The force would be transmitted through levers to the brake shoes, per the prototype. The steam line for the tender brakes is plumbed into the steam line for the engine brakes so the same valve controls both the engine and tender brakes.

I bought the small brake shoes from Allen Models/Alley Supply. They are cast iron brake shoes. I bought enough for the three tenders plus three extra.

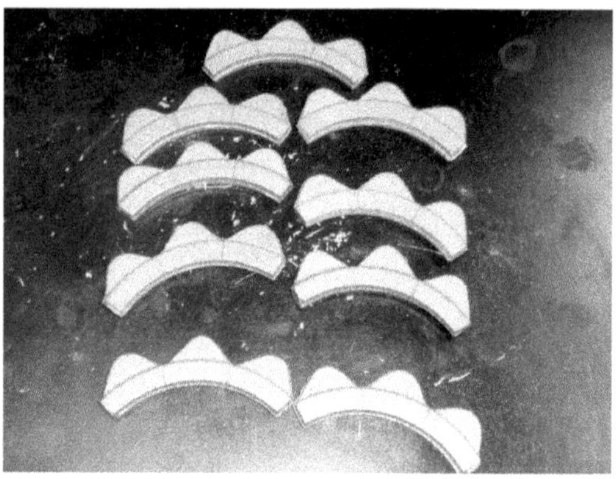

Left: **Figure 196. I bought the miniature brake shoes from Allen Models.** *Right:* **Figure 197. The attachment hole has been drilled and the rubber pad has been glued on.**

I glued/bonded a rubber brake pad (I bought a couple of vacuum cleaner belts and de-glossed the rubber to get it to stick). As the rubber wears, I can easily replace it instead of replacing the whole shoe. Cast iron shoes will also wear the wheels and when figuring out how much force that needs to be applied to the shoes, I found that cast iron shoes need a force around 80 percent of the weight that is on the wheels. The lower friction composite shoes need a force around 45 percent of the weight on the wheels and don't wear the wheels. The rubber is closer to a composite brake shoe so for my calculations, I calculated the braking force to be around 40 percent of the weight and added extra holes to the live/dead lever on the brake cylinder to make other adjustments if needed. If the brakes are set too hard, the wheels will slide.

To figure out what lever ratios I needed, I sketched them out. The 2-4-0's tender is around 160–170 pounds and with an engineer on the tender, the total weight could be 350–400 pounds. Forty percent of 350 pounds is 140 pounds of braking force and 40 percent of 400 pounds is 160 pounds of braking force. Since the drivers will slide when full boiler pressure (130 psi) is applied to the brake cylinder, I based the calculation for the tender on 100 psi, 40 pounds of force. (It is possible that the pressure may be lower than this due to the steam condensing.)

The brake rigging is a series of levers that can increase the force (decrease travel distance), decrease force (increase travel distance) or keep the force (and the travel distance) the same.

The piston rod of the brake cylinder attaches to the live lever (gets the force directly from the piston). A tie rod connects the live lever to the dead lever (gets the force indirectly from the piston). The dead lever is connected to the other end of the cylinder. With this arrangement, the net force on the cylinder mountings is zero pounds. Adjusting the lengths of the placement of the holes on the live and dead lever changes the amount of force that will be passed to each truck.

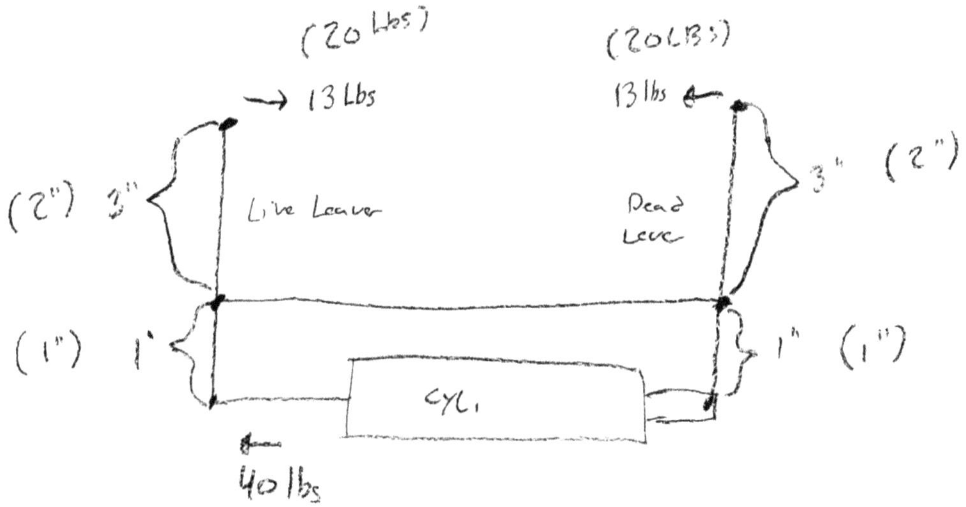

Drawing 45. With the brake cylinder giving 40 pounds of force, these lengths of the live and dead levers will give 13 pounds of force or 20 pounds of force to each truck.

Drawing 45 shows the variations of lever lengths to pass 13 pounds or 20 pounds of force to the trucks with a cylinder force of 40 pounds.

Tie rods connect the live and dead lever at the cylinder to the live levers (which get the direct force) on each truck. The "middle" hole of the live (and dead) levers connect to the brake beam and the bottom connects to a tie rod that transmits the force to the dead lever (which indirectly gets the force) to apply the braking force to the other axle.

Drawing 46 shows the sketch I used for the 2-4-0. With these levers on the truck, one pound of force on the live lever gives six pounds of braking force.

My first guess with lever ratios gave me 73 pounds per axle, or 292 pounds of force total. My second guess gave me 48 pounds per axle, or 192 pounds of force total. My third round of levers gave me 39 pounds per axle, 156 pounds of force total.

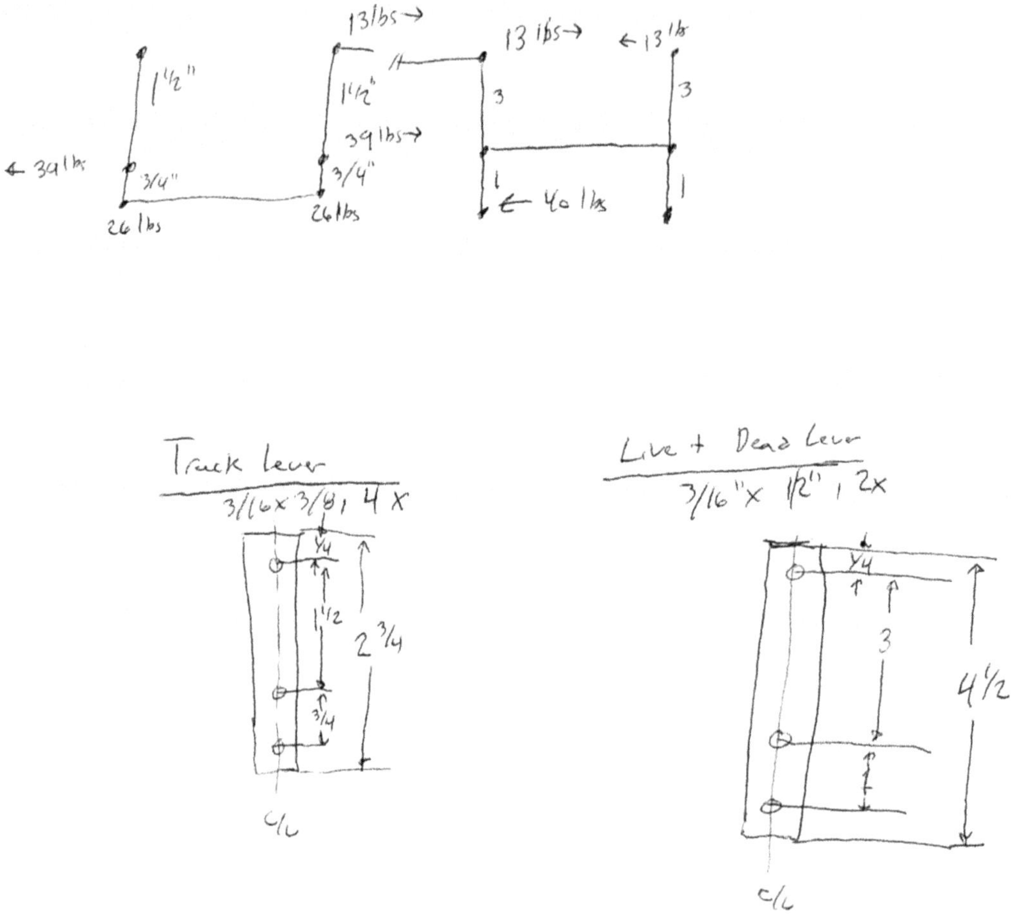

Drawing 46. With 13 pounds of force going to each truck, the live and dead levers on each truck are set up to deliver 39 pounds of force to each axle.

For the 2-8-2's and 4-6-0's tender, I left the set-up of the truck levers (the live and dead levers on the truck) the same as they were for the 2-4-0. I adjusted the force that is applied by adjusting the live and dead levers on the brake cylinder and used the same brake cylinder on all three tenders.

For the brake cylinder I bought a ¾" diameter, 2" stroke single-acting rear pivot mount cylinder from AutomationDirect (http://www.automationdirect.com/adc/Shopping/Catalog/Pneumatic_Components/Pneumatic_Air_Cylinders/Round_Body_Air_Cylinders_%28A-Series%29/Single-Acting_-z-_Spring_Return/A12020SP).

While these are air cylinders, I have had good experience using them in this type of situation for steam brakes. The cylinder is generally a ways away from the steam valve and the steam will condense in the line, filling the cylinder with water. If the brakes are applied and released multiple times in a short order where all the steam is allowed to bleed off, the plumbing will start heating up and the steam won't condense until farther in the line and possibly getting to the cylinder. If there is a leak in the line, the condensation will escape and steam can reach that point. The steam will cook the cylinder and then it will need to be replaced; which is pretty easy using this set-up.

With using this type in a daily basis (4–5 hours a day), the cylinder closest to the steam valve lasts one or two months of straight running. The cylinder farther away from the valve lasted more than a year of 120–150 days of operation per year.

At 100 psi this cylinder gives 40 pounds of force and the pipe fitting is ⅛ NPT. To hang the cylinder to the tender frame, I made a holder out of ⅛" × ¾" steel. I bent the ends over and drilled a ½" diameter hole in one end (had to egg it out slightly to accept the ½–20 threads on a slight angle) and slotted the other end. I used a ½–20 nut to hold the cylinder to its frame and made a plate to connect the cylinder frame to the bottom of the tender frame.

Figure 198. The brake cylinder is in its frame with the live and dead levers attached. The cylinder frame will be bolted to the tender frame.

The live lever is attached to the piston rod of the cylinder and the dead lever (I made both from 3⁄16" × ½" steel) is attached to the dead/stationary end of the cylinder.

Figure 199. Top view of the brake cylinder showing the connections of the live (on the left) and dead levers.

All of the force is applied to the cylinder and not to any of the cylinder holder. The piston applies 40 pounds of force to the live lever. With the pivot being 1" away from the piston rod and the connection to the truck being 3" away from pivot, 13 pounds of force will be transmitted to the lever on the truck. To give adjustment capability, I drilled two extra pivot holes, one on either side of what I want, so I can make the brakes stronger or weaker as needed.

The levers on the truck are designed to magnify the 13 pounds of incoming force to 39 pounds on each brake beam. One lever on the truck will be connected to the live/dead lever from the cylinder and the second lever on the truck will be connected (as a pivot point) to the truck; typically to the bolster. With the hole set up with 1.5" between two holes and 0.75" from the "middle" hole to the third hole, each pound of force applied to the top of the lever is amplified to three pounds of force to the "middle hole"/brake beam.

After measuring the trucks, I decided to make the brake hangers have 2" centerlines. This would allow the foundation rigging to still be an ample amount above the rail head.

Figure 200. The brake hangers and shoes are mounted on the truck.

The prototype hangers would typically have the brake shoe connection closer to the pivot than where the brake beam connects. Doing it that way will also amplify the force applied to the brake shoe. Since I don't need to amplify the force delivered to the brake beam, the brake beam connects to the brake hanger at the same connection as the brake shoe.

I made the brake beams out of ½" square steel and made my own clevises out of ½" square steel. Once most of the parts were made, I started fitting things up and making the pieces that set on top of the bolster for the upper connection of the brake hangers.

For all of the connection rods, I measured their lengths as I was assembling the pieces. To make the connection levers from the cylinder dead and live levers to the truck live levers I had the trucks and cylinder mounted on the tender frame and set the brakes so the tender live lever would need to move about 1.25" to apply the brakes and then measured how long the tie rods needed to be. Due to the ratio of the cylinder live and

Figure 201. This end view shows the dead lever, brake beam and connections for the brake hangers.

dead levers, this meant that the brakes would be applied before the piston reached its 2" max travel.

Figure 202. The brake cylinder has not yet been plumbed up but the tie rods from the cylinder dead and live lever are seen to the live levers on the trucks.

Note: I used 10–32 socket head cap screws throughout the brake rigging. Where I couldn't use a nylon lock nut, I used Loctite on the threads. The bolts holding the brake shoes to the brake beams and brake hangers were Loctited in place, ran in snug, put on the truck, brake shoes adjusted and then tightened down the rest of the way.

The tender brakes are plumbed into the brake line to the engine brake cylinder so that way one brake valve controls both the engine and tender brakes. For the flexible line between the engine and tender I used a 12" long stainless steel braided hose (McMaster-Carr #4468k81) which has 7⁄16–20 female threads which is the thread for a ¼" tube flare fitting (http://www.mcmaster.com/#braided-stainless-steel-hose/=zzz1vx).

When I ran the engine and tried the brakes, the brakes worked with well. With the 2-4-0 having around 270 pounds on the drivers and now the tender (which can weigh 350–400 pounds) having brakes, the braking capacity has more than doubled. I don't think I am getting quite all of the braking capacity out of the tender as I can. Part of that could be that the pressure in the cylinder is lower than the steam pressure due to the steam condensing. Part of it could be that the rubber needs a higher percentage of force than 40 percent of the weight. Part of it could be a loss of energy through the levers. Whatever the reason is, I am hesitant to make the brakes much stronger, as if they are too strong for the weight (for example if there is a smaller engineer on the tender) or it is slippery, the wheels will slide and the braking power will be greatly reduced. The 2-4-0 can now stop an 1,100-pound train running down a straight 3 percent grade around 3.7 mph in about 8' and can stop the same train running around 5.8 mph in about 15' without sliding the wheels. This is a huge improvement from just the engine having brakes.

I have installed tender brakes on the 2-8-2 and 4-6-0. The brake hanger connections are different between all three engines as they each have a different style of truck. The tender live and dead levers are the same length between all three engines. I modified the brake cylinder live and dead levers to get the desired force (around 40 percent to 50 percent of the tender weight) for braking.

Tender Brakes versus Car Brakes

Several people have emailed me asking why I would put brakes on the tenders instead of on the riding cars as the riding cars can weigh more than the tenders.

The brakes are set so when fully applied, the wheels on the car won't slide when the car is empty. If a flat car weighs 100 pounds empty, the brakes are set so the 100-pound car won't slide even if it can weigh 500+ pounds when carrying a couple of adults. (A large riding car may weigh around 300 pounds empty but could carry four adults and weigh 1,100 pounds loaded.)

The tenders for the 4-6-0 and 2-4-0 weigh around 150 pounds with water and coal. With an engineer on the tender, the tender weight could be around 350+ pounds. The 2-8-2's tender weighs around 300 pounds with coal and water and can weigh around 500 pounds with an engineer on it. It would take 3+ flat cars to equal the braking force of the 2-4-0's and 4-6-0's tender and five flat cars to equal the braking force of the 2-8-2's tender.

With tender brakes, steam from the boiler can be used and brakes need to be set up for the tender. (Setting up the brakes can be 10–15 hours.)

For car brakes, an air compressor, tank and new brake valve would be needed. Alternatively, an ejector, valve and piping would be needed and each individual car would need a cylinder and foundation brake rigging.

With brakes on the tender, you get more bang for your buck and time.

Some arguments for car brakes use comparisons to the full-size; when a full-size engine's brakes are applied, the cars will just pile up behind the engine. Full-size cars are a lot heavier and have a lot more momentum than the miniature. For 1:8 scale, the mass/volume is $1/512$ (0.195%, $1/8^3$). Momentum is mass times velocity. So if a full-size flat car and 1:8 scale flat car are traveling at the same speed, the 1:8 scale flat car will have 0.195 percent the momentum of the full-size. If the 1:8 scale car was traveling at a scale speed (one-eighth speed) as the full-size car, then the 1:8 scale car would have 0.024 percent the momentum of the full-size. If a 120-pound car is carrying two adults who weigh 250 pounds each, the car will weigh 620 pounds. When compared to a full-size, loaded freight car in the U.S. (260,000 pounds+), the loaded 1:8 scale car is only 2.4 percent of the mass of the full-size.

One drawback with having brakes on every car on 1:8 scale trains is that it is possible to stop too fast; the momentum will cause the passengers to shift on their seats, have the seats move on the car or for the passengers to fall forward out of their seats if the train stops too quickly.

Using the Prototype Car/Tender Brakes versus Cylinders on Each Truck

Different vendors simplify the brake rigging by installing brake cylinders between the brake beams on each truck. This means that there are fewer parts to make and the only thing that needs to be installed on the car body is the brake pipe line. As the shoes wear, the brake cylinders will travel more due to the thinner brake shoes.

On the tender trucks, I didn't have a lot of room between the axles to install a brake cylinder and in the event of a derailment it would be very possible to damage the brake cylinder that is mounted on the truck.

With using the prototype design, the brakes are also self-adjusting as they wear and I could use a single size of a cylinder between all three tenders and modify the levers to get the needed force. If the brake cylinders are on the truck, I would need to have larger cylinders to get more braking force. Also, with having a single cylinder on the tender, I had less plumbing to do.

Sanders on the 2-4-0

Note: The information in this section can be adapted to any steam, diesel, electric or gas locomotive of any gauge.

The 2-4-0 is generally worked up closer to her maximum capacity more than the other two engines and has a tendency to start to slip when she would be around 80 percent of her capacity. This would wear the rails and put some nice, clean divots in the rail.

Figure 203 shows slip marks on the rail; the marks are still there after one year of using sanders on the 2-4-0. To get the traction I needed (she has plenty of power), I would manually sand the track. I have thought about adding sanders to her over the past couple of years. Sand is a great tool to use (but not overly) for getting more traction. In some old books it says that using sand on wet rail will give the same traction as on dry rail. Only little shots of sand are needed. If too much sand is applied, all the cars will be on a lot of sand and that will add more resistance.

Figure 203. When the drivers slip, they leave small scallops in the rails. These were still present one year after using sand.

To seal the sand dome (as it was meant for a cosmetic piece), I bought 4" diameter shim stock (0.010" thick, used three pieces) and bolted them to the sand dome bracket and sealed the bottom.

***Above, left:* Figure 204. The sand dome was bolted to the top of the boiler as a cosmetic piece. It was not originally designed to hold sand. *Above, right:* Figure 205. The sand dome was taken off the boiler and the piece of round shim stock would be applied to seal off the bottom. *Left:* Figure 206. The shim stock was bolted to the bracket that held the sand dome to the boiler and then sealed.**

I drilled and tapped into the sides of the dome for ⅛ NPT fitting and used 45-degree ¼" compression fitting (fig. 207). For the sand trap I used an ⅛ NPT tee (fig. 208) and plumbed it so the sand would drop in front of the front driver (fig. 209).

***Above, left:* Figure 207. I drilled and taped a ⅛ NPT hole into the side of the bottom of the sand dome and used a 45-degree (⅛ NPT to ¼" compression) fitting to connect the delivery pipe to the sand trap. *Above right:* Figure 208. The sand trap is made from a ⅛ NPT tee with fittings for the sand and air lines. *Right:* Figure 209. The pipe to the drivers went between the boiler and running board and in front of the first driver. Since the engine runs predominately in forwards, I wanted to have sand for all drivers when running forwards.**

I kept the line at least 1" above the rail for easy access for cleaning. I ran the air lines from the sand trap to under the right rear corner of the cab.

***Left:* Figure 210. The air line on the left side ran above the running board and then ducked under the cab before cutting over to the right side. *Right:* Figure 211. The air line on the right ducked under the step in the running board**

Figure 212. The air lines from both sides joined under the right rear part of the cab. The flexible line from the tender was piped in.

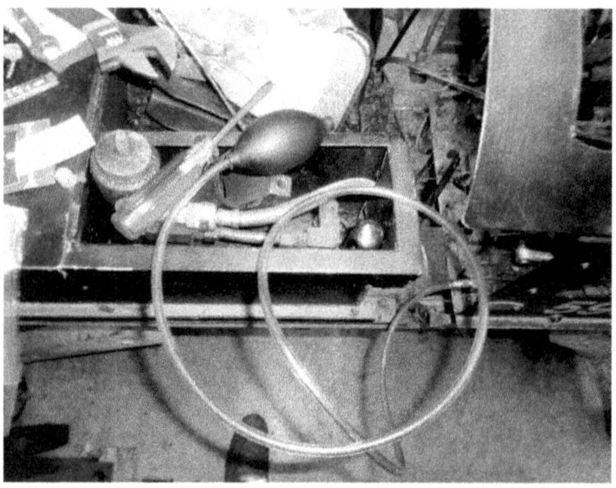

Figure 213. The air bulb will give small puffs of air which will knock the sand out of the sand trap (tee) where it will then fall down (by gravity) the pipe onto the rails.

To force the sand out of the sand trap, I bought (on Amazon) a large air pump bulb that is used on blood pressure cuffs as it doesn't take a lot of air pressure to force the sand out of the trap. I did not use the small diameter outlet valve but inserted a piece of ¼" OD rubber tube into the bulb and connected that to the air line connection on the engine.

Squeezing on the bulb delivers enough air to force the sand through the sand trap and onto the rail head. Here is a video of the sanders working: https://www.youtube.com/watch?v=opRCSRGcTNs.

I first tried to use play sand, but that was too fine and compacted too much. Slightly coarser sand-blasting sand works well. The sanders are a great tool and really help the engine be able to pull her maximum load. On my track she is rated to pull a 1,540-pound train up the 3 percent grade around the 52' radius curve and she would start to slip with a train weight around 1,200 pounds. With the sanders I can give a couple of quick shots of sand to help her hold the rail (not on spinning drivers as that could cause one side to grab while the other side is still spinning) to pull well over 1,540 pounds up the grade.

I am not too worried about the sand ruining bearings. On the 15" gauge 4-4-0 and 4-4-2, the sand has not gotten into the bearings. On the coal burners it appears that there is more wear in the bearings closer to the firebox and ashpan as they get exposed to the abrasive ash. This has been seen on full-size engines as well.

I am also not concerned about the sand causing undue wear on the drivers and wearing prematurely hollow. The 15" gauge 4-4-0, steel tires on steel rail, runs around 1,000+ miles a year and tires didn't wear prematurely hollow. The 15" gauge 4-4-2, steel wheels on steel rail, ran 3,500–4,000 miles a year in Traverse City. Her drivers were worn a bit after seven seasons (24,500–28,000 miles), but nothing more than what is expected for the miles she ran.

To Use Sand or Not to Use Sand

I have heard that sand should not be used as it will wear the rail faster. Rail wear will happen no matter what. On curves, the flanges will wear the outside rail of the curve. Cast iron and steel are harder than aluminum rail and over time they will wear the rail head. I have also heard that sand will get into the bearings and ruin the bearings. The dust from ash can also get into the bearings and on several engines (miniature to full-size) I know have more wear in the crank pins at the crank pins that are most exposed to ash and not the ones near the sander.

After having used the sanders for a year I don't see any extra rail wear. The most sanded rail is 0.832" tall and the new, never used rail is 0.831–0.834" tall so I haven't lost any rail height due to the use (not abuse) of sand.

Figure 203 is of rail that was damaged by slipping. This rail is at the steepest part of the grade and has seen the most amount of sand use. The rail on the curves (where sand was used and where sand was never used) shows uniform wear from the flanges. The wheels (lead truck, drivers, trailing truck, tender and car wheels, a mixture of steel and cast iron wheels) do not show any more than average to moderate wear. I don't see any more wear than I did before I installed the sanders. All of the bearings have been running cool, no major wear in the side rod bearings and none of the rods or bearing surfaces has had any sand show up on them. I have minimized the wheel slip and damaging the rails due to wheel slip.

I have added sanders to the 4-6-0 and 2-8-2 using the same technique, but I mounted the sand trap just below the sand dome as it was easier that way. Sealing up the sand dome had to be done differently due to the different dome designs. The sanders work well on those engines too.

My Experiences in the Hobby

When I was little I remember watching a Disney cartoon where Donald Duck had his own miniature steam railroad and I wanted to have a small steam engine like that. I was probably around four at the time and was fascinated by trains. I liked going down to the Clinch Park Zoo to ride on the Spirit of Traverse City there. The old depot in town was then a bank and the owner had on display an old steam engine from France, as well as a boxcar and a caboose; I liked to have Dad take me down there to look at them. Sometimes when we would hear the local freight blowing for a crossing, Dad would take us out and we would go chase the train.

One day a week Dad would be home with me and my sister while our mom worked. Dad would usually take us on different trips around the area. One time we drove down to Cadillac to find the Shay that is on display in one of the parks. Another time we tried to find a miniature railroad that was north of Cadillac. It took several trips but Dad found a guy who knew Bill Fitt and the railroad. Bill happily showed us his engines and cars and we saw mostly the tracks that were in the yard. Bill encouraged my dad to get started in the hobby, buy some machinery and start "making scrap" and build and/or buy a live steam engine. Being so young when my dad got started in the live steam hobby, I consider myself to have grown up in the hobby.

The Shay

Around the fall of 1989 Dad saw an ad in *Modeltec* for a 1.5" scale, 7.5" gauge Shay that was for sale in Washington due to the owner's failing health. He inquired about the engine and ended up buying it. Our backyard was pretty small and a track would require curves down to a 20-foot radius and he wanted an engine that would be able to handle the tight curves. The Shay arrived during the winter of 1989/90 and I remember my parents dragging the Shay (in a crate) on a sled through the snow to get it to the back door and inside. For a short while, the Shay was kept in the basement, but the smell of kerosene got to be too much so it was moved next to the rear door.

The Shay was built by Dutch Boysen of Yakima, Washington. He had built several other engines and the Shay was his last one. He used William Harris's plans for the Mich-Cal Shay as the basis for the engine. Instead of the Tee boiler, he built a horizontal Briggs boiler and burned kerosene in a pot burner. I heard that he modeled the engine after a Shay that had operated near Yakima. To carry more water, he had made a logging car that carried two cylindrical tanks made from PVC pipe. Figure 214, taken

Figure 214. Dad bought this little Shay locomotive in 1989 from Dutch Boysen (courtesy L. Lee Frost).

around 1991, shows me running the Shay at Bill Fitt's Meauwataka and Wildwood railroad.

During the first year that we owned the Shay, we did not have any track on the ground. During the summer, Dad kept the Shay on a stand in the garage. To run the engine, he lifted the engine off the stand and supported it using 6 × 6s under the front and rear beams. He disconnected the square drive shaft to the rear truck and kept the front truck connected to use the axle pump. Dad taught me early on the importance of having water in the sight glass and showed me how to operate it and take care of it. This was a learning curve for both of us.

The Shay was not without her share of problems. Early on Dad found out that the blower liked to plug up so he would not have much draft and that the axle pump did not like to work all the time and did not keep up with water consumption. More than once he had to push her back from partway around Bill's track. Later on when I was old enough, I got to help push her back as well.

The Shay was a slow engine. We thought that she was supposed to run that slowly. After all, she only had 4" wheels, small cylinders and was a geared engine. In 1995 we met Rodney, who had a 1.5" scale class A Climax and we were really impressed at how much faster his cylinders moved compared to ours. He looked at the Shay and saw that there was a lot of slop in the eccentric straps and told us to look at the valve gear. We replaced the eccentrics and made new straps. (Later on, we found out that the offset for the eccentrics from the plans is not what it should have been. This did not help timing the engine as the valve could not give even port openings and have the engine run at all.)

Rebuilding the valve gear did help some. But because we were still learning a lot, we didn't know really everything to look for and change to help her run better. Towards the end of our ownership of the Shay, I realized that the cylinders were not being fed enough oil. The Shay had a displacement lubricator just before the throttle. The throttle was well lubricated, but there was not enough oil making it to the valves and pistons. Once we increased the oil feed, the engine ran quite a bit faster and smoother. The problem was that the oil reservoir emptied rather quickly.

The 0-4-0/2-4-0

Around 1990 Dad decided that he also wanted to build an engine. We were going to have a separate shop building and he was starting to buy some machines. He had also bought some track and started laying track around this time. After some research, Dad decided to build the 0-4-0 that was (and still is) offered by Railroad Supply.

Because I was only six when Dad started to build the 0-4-0, I did not do too much in the early stages. As I got older and could do more things, Dad let me help measure parts, drill holes, help to fit up the valve gear castings and he taught me how to use the mill and lathe. He also taught me a little bit of welding. Because both of us would not be comfortable with welding a boiler (we still aren't), we had Ed Perry build the boiler for the 0-4-0 and later our 4-6-0.

The engine first ran in May of 1996 so she was numbered 596. She ran at the very last Tri-State Locomotive Club meet at the Meauwataka and Wildwood railroad. The engine was not complete: headlight, bell, cab were missing and the tender tank leaked, but we ran her. Compared to the Shay, this engine was a very nice size and was easier to run. However, we still had our share of problems.

The axle pump did not like to always work. We had the hand pump set up to pump through the axle pump which saved us multiple times when the pump would seize up. As it turned out, one of the check valve balls was too small and once we replaced that, the pump worked as it should. By 1998 we had disconnected the axle pump. For us, it acted like a hydraulic brake and you could feel each time it pumped water into the boiler. Also, the cold water would drop the pressure quite a bit. The boiler only holds two gallons at running level. We had tried injectors before, but had a hard time finding one that would work above 100 psi. Then we tried SuperScale's Chicago injector. This worked great and now on all of our engines we use two injectors (SuperScale's Chicago and Economy injectors) and have a hand pump in the tender which is used for hydro tests each year. For us, injectors were the way to go and they have worked well over the years and at varying pressures.

We also had a hard time keeping up steam pressure. The safeties were set at 125 psi and if we would drop down to 60 psi, we thought we were doing great. We had to learn how to fire coal, how to run the engine more efficiently and also needed to learn the proper ratio of grate bar size to the air opening and install an arch.

In July 1996 we first visited the White Creek Railroad in Cedar Springs, Michigan. We were very impressed by the track, which at that time was only around two miles in length. Figure 93 was taken at the White Creek Railroad. My grandfather, my dad's father, was running the 0-4-0 and I am riding behind him. My parents had the picture enlarged and framed with the title "Ageless Pursuit."

There were times when we had to push the 0-4-0 back due to the flues being clogged and not being able to create steam, or something broke or a pin came loose. Typically after each visit, we would need to repair or improve something on the engine and then go back and try it out. Sometimes it worked, other times not so much. A lot of our learning was through trial and error.

Over the years different parts have been added or modified. One time when my brother was running her, she derailed over a switch (there were stones in the frog) and the cylinder cocks hit the rail. One of them hit with enough force that it broke off a piece of the cylinder casting. In order for this to not happen again, I redesigned the cylinder cocks so that the cylinder cock would be sacrificed and not the casting and then added a guard under the cylinder cocks. For the most part, the guard protected the cylinder cocks.

In 2001, when I was running the 0-4-0, she split a switch and tore the cylinder cocks off one side of the engine. Dad was running the 2-8-2 and was behind me. Once we got the 0-4-0 back on the track, we dumped the fire and pushed her clear into the siding. Dad then bought the 2-8-2 and his train beside me and we lifted the 0-4-0 onto a flat car, coupled the engine to the 0-4-0's train, and took her back to the yard on a flat car. After that I designed the lead truck and modified her into a 2-4-0.

Figure 215. Don Frost is running the 2-4-0 (the gondola behind the tender is from Bill Fitt's track) at the track at my house in Pennsylvania. ***Bottom:*** **Figure 216. Don Frost is running the 2-4-0 at the White Creek Railroad in the summer of 2002. The 2-4-0 is pulling a nine-car train, though only five are visible.**

Also during this time, we made new grates to let more air to the fire and installed a stainless steel arch which improved her steaming qualities and had a new, larger tender tank made so the water and coal supply would be increased.

Before these improvements, we were only able to pull six cars around the White Creek Railroad track, as she would not be able to produce enough steam for more. After the new grates, arch and rebuilding the valve gear, we were able to pull nine cars around most of the track and she was steaming beautifully.

The Mikado

In the fall of 1996, the 0-4-0 was running but was not yet complete or fine-tuned, Dad and I began talking about what we wanted to build next. Dad wanted a bigger engine and we decided on a 2-8-2. Railroad Supply had the 2-8-2 available, but then the production of the castings changed hands several times. When we contacted the company who had the castings, Dad got an almost full set of castings at a good discount. Later we found out that these castings were not the best we had seen (some of them broke) and were probably the leftovers that were lying around during the time that the Mikado production kept changing hands.

We started machining parts in late 1996. The frame went together without too much problem. We made a few mistakes, but corrected them by adjusting the parts that would attach to those holes so the overall line up would be correct.

During one of our trips to the White Creek Railroad in 1997 we met Terry Weaver of Coopersville, Michigan. He is a machinist and fabricator/welder by trade and was interested in the live steam hobby. He ran the 0-4-0 and really enjoyed it. We came to an agreement that we would trade the Shay (which was in better shape than when we bought it, but still needed fine tuning) for him doing the heavy machine work on the cylinder castings and building the boiler for the Mikado.

By New Year's Eve of 1998 we had the Mikado running on air for the first time. The engine was stiff but we just passed it off as the bearings having close tolerances on the pins. (Later on we found out that this was not the case.) In April 1999 we picked up the boiler from Terry and the Mikado first steamed in the end of May 1999. We numbered her 1555 as I was 15 and Dad was 55 when we finished her (see fig. 3).

One of the first problems we had was that the engine rode hard and the trucks did not want to swing their full amount. Also, the lead truck was bearing hard against the left rail. We took out the lead truck and the pivot point was not in center with the truck. After adjusting the pivot hole in the truck frame, the truck tracked properly. For the engine riding hard, it appeared that the equalizing (spring rigging) linkage to the trucks was not permitting the trucks to swing easily.

On the prototype, engines aimed to have a three-point suspension. The lead axle(s) would be equalized using a cross equalizer (so the springs on the right lead axle(s) would be in connection with the springs on the left) and they would be tied in with the single-axle lead truck. (For the two-axle lead trucks, most engines did not have the equalizing tied into the lead truck. This would have the lead truck set the height of the pilot.) This would control how high the pilot of the engine sat. Then the rear axles of each side would be equalized on their respective sides only; the right rear springs would be equalized with each other and the left rear springs would be equalized with each other. The right equalizing would control how high the right side of the engine sat and the left equalizing would control how high the left side sat.

We were having problems with the equalizing not functioning properly. The heart links of the lead truck (used to center the lead truck) would bind up (later we found out that the nuts on the bolts were tightening themselves up and binding. I guess we should have used a locknut or a pin with cotter pins to keep it from doing this) and the homemade springs on the trucks were not working well. The engine also tended to be slippery. A fellow live steamer suggested that we disconnect the equalizing to the trucks and then adjust the springs and spring it as an 0-8-0. Then put separate springs on the trucks and have them carry enough weight so they won't derail. We re-sprung the engine and she rode better and took the curves more easily. However, others saw what we did and questioned us on why we did that and told us why it should not work well. As with most things in this hobby, everyone has their own opinions and ways of doing things. Sometimes they work for one group but they don't work for others. For our 2-8-2 it worked decently for the most part. In 2014 I made a new swing bolster for the lead truck as the truck was not doing a good job of leading the engine around curves and the lead drivers would climb the rail. At the time when we changed the spring rigging, we really didn't understand how it was supposed to work and followed the advice of a few fellow live steamers who had more experience than we had.

After running the engine for a while we noticed that our flanges were getting thin from rubbing on something on the frame. There were screws that stuck out and as the wheels shifted in the frames, the screws would wear down the flanges. We later discovered that when we machined the driving boxes, we made them to the dimensions for a 7.25" gauge and not 7.5" gauge so we gave the axles a lot of lateral play. In order to save the wheels, we added steel tires to the hubs.

During a Tri-State Locomotive Club meet at the White Creek Railroad, one of our crank pins came out of the wheel. Dad and I just finished going down a steep grade with a train of about eight cars when the left main crank pin came out. We disassembled the rods on the left side and then had to push the train back up the hill and into the yard and then push the engine onto the steaming bay. Upon inspection we found that the rod connecting the L2 pin to the L3 pin was 1⁄32" longer than the axle centerlines. This caused the engine to bind and fight itself until the pin came out and was a big reason why the engine was stiff when we ran her on air. The next day I installed the new side rod and she ran well and rolled a lot better. As time went on, I made new side rods for the engine and fabricated them out of steel instead of aluminum.

For the most part, once we corrected these problems that showed up in the first year, she ran well and we generally did not need to do any major repairs to her. It is usually at a time like this that the engine will decide to throw you a curve ball. Dad was running the Mikado and we were just starting up M.T. trestle at the White Creek Railroad when all of a sudden the engine locked up and we came to a stop. I got off and walked along the ground and watched Dad try to go backwards and then forwards, but the engine would lock up when the left piston was near front dead center.

Figure 217 shows the approximate location where this happened. We are in the habit of carrying tool boxes, extra valves, coal, water, crowbar, etc., on the train for times like this. I took off the left front cylinder cover (standing on the supports of the trestle) and found that the nut locking the piston to piston rod had fallen off and became jammed between the piston and the front cylinder head. Once I removed the nut, the Dad ran

Figure 217. Dad is running the 2-8-2 up M.T. trestle on the White Creek Railroad around the summer of 2004. It was around this spot that the piston rod nut on the left piston came off and I had to climb up the side of the trestle to remove the front cylinder head.

the engine off the trestle and back to the steaming bay. Once we were home, I replaced the nut and peened over the end of the piston rod so the nut could not fall off again. I then did this to all of our engines.

During different trips to the White Creek Railroad we would invite some family (my dad's family is from Hastings, Michigan) to come up and join us. Figure 218 shows my cousin's daughter running the Mikado and I am sitting right behind her. In the spring of 2002 we had a birthday party for my cousin's five-year-old son and we double-headed the 2-8-2 and the 4-6-0 (fig. 219).

Figure 218. My cousin's daughter had a chance at the throttle in the summer of 2002 (courtesy Donald S. Frost).

Another problem that we had is when we pressed the wheels onto the axles, they did not press on straight. This meant that as the wheels turned, the crank pins were wandering and did not have a constant centerline. Dad made new axle sets and we made the journal boxes correctly, added tires to the drivers right from the start and made sure they were pressed on straight. We then had to fabricate new side rods to match the new centerlines of the crank pins. I did most of the designing and fabricating when I was in Traverse City so Dad really did not do much of it. For the new side rods, I designed them, made plans and showed dad how to make the parts so he could fabricate the new rods. Once we corrected that problem, she ran even better.

Figure 219. In the spring of 2002 we double-headed the 4-6-0 and 2-8-2 for a birthday party for my cousin's son (courtesy L. Lee Frost).

Figure 220. The last time we were at the White Creek Railroad was in the summer of 2010. Dad is pulling out of Cornertown.

The Ten-Wheeler

During the summer of 2000 we wanted to build another engine. After looking at different engines, we decided to build the Allen Models ten-wheeler. Dad wanted an engine that was not as small or fussy like the 2-4-0 (due to the small boiler she is a busy engine with feeding coal into the firebox or adding water to the boiler) and something smaller than the Mikado so he could move it around and re-rail it by himself relatively easily. After looking at the engines we decided that the 4-6-0 would be a good medium-sized engine and it looked like it was relatively easy to build. Construction was started in September 2000, Ed Perry build the boiler for it and she first steamed in May 2001. Dad, who is a retired OB-GYN, said that the ten-wheeler was a "typical pregnancy." She was built in memory of my dad's father, Howard Alton Frost, who was born in 1908 and loved the color red.

Figure 221. Dad is running the 4-6-0 on the short track (about 70') we had in Traverse City, Michigan, in the late winter around 2005.

During construction we decided that instead of using the Stephenson valve gear as the plans called for, we would use the Walschaert valve gear. I used one of Dad's old railroad books on valve gear and designed and fabricated all of the parts for the valve gear. Some of the parts were a bit bulky because that was the size stock we had. After the engine had been running for several years I decided it was time to trim down the parts. During the construction and fine tuning of the valve gear, I only need to make minor changes to a few parts to have it function properly.

During the first run of the 4-6-0 at the White Creek Railroad, the only thing that caused a problem was the lead truck. It kept derailing because it was not carrying enough weight. Once we got more weight on the truck, she ran very well.

Figure 222 was taken in 2001 at the White Creek Railroad. On the stack is our home-made spark arrestor made from aluminum screen that would be used on a screen door. It is held to the stack by using a hose clamp. The 4-6-0 and 2-4-0 can throw sparks out their stacks and when it is dry we will use this type of a spark arrestor. We also then added a ¼" square mesh (in 2016 it was replaced with $^3/_{16}$" square mesh) screen inside the smoke box of the 4-6-0. It wraps around the petticoat and extends down to the floor of the smokebox.

Figure 222. The 4-6-0 was completed in May 2001 and this picture was taken at the White Creek Railroad in the summer of 2001 (courtesy L. Lee Frost).

After the engine had been running for about a year, we decided to convert her over to propane to be able to run her during the dry summer months. For each of our engines the boiler was secured to the frame and then the grates, their holders and ashpan were assembled with the boiler in place. This way we can get them out with the boiler still attached to the frame. For the 4-6-0 we bought propane burners from LocoParts. Once the ashpan was out, we made a holder for the burners and were able to install the burners while the boiler was still on the frame. To cut back the draft, we replaced the ¼" screen mesh in the smokebox with a sheet metal sleeve and lowered the petticoat. We also made a propane car out of one of flat cars. We can now convert the 4-6-0 back and forth between coal and propane in about an hour.

General Improvements

Over time we have been improving different things on our engines. During construction of the engines we decided to use a ball valve for the throttle and place it in the smokebox in front of the smokestack. We have had very good experience with ball valves and they have lasted very well in front of the smokestack. We have not needed to replace any throttle valves to date.

I have rebuilt the valve gear on the engines multiple times and each time I have adjusted the length of the eccentric rod, the setting of the eccentric, the radius on the link and have replaced bushings. For the most part, with each adjustment the engine ran better. On our 2-4-0 we can use all four forward notches in the Johnson bar. On the 2-8-2 we can use all seven for forward, and with the throttle wide open we can run in neutral. On the 4-6-0 we can use all six notches in forward. For reverse we don't have as many notches as we don't run in reverse too much.

We have added steel tires to all of our drivers, which have also helped adhesion. After multiple attempts with trying to make an arch and a holder for the firebox (to lengthen the path of the gases and promote better combustion) I was finally able to find a way that worked. It made the greatest difference on the 2-4-0 and the flues did not plug up nearly as quickly. We went from needing to clean them every 2.5–3 hours to running for five hours. After five hours, the flues were somewhat dirty but the engine was still steaming well.

When we built the engines, only the 2-8-2 had an equalized brake rigging. In 2006 I built equalized brake rigging for both the 2-4-0 and the 4-6-0. With this, each driver gets an equal amount of braking force and it has helped reduce the wear of one brake shoe over another.

Today we have three engines that run very well. It certainly did not start out that way. We had a lot to learn and many improvements were done by trial and error. We learned a lot about how steam engines operate and a lot about their design and we are still learning and doing improvements.

Leaving Home

I graduated from high school in 2002 and attended Michigan Technological University. Before I left for college, Dad and I brought all three of our engines together to get a picture of them.

During the summers while I was in college, my dad and I started working on the 3" scale 4-4-2, the Spirit of Traverse City, and led a complete overhaul of the locomotive.

Figure 223. In 2002 we lined up all three of the engines for a picture.

Figure 224. During a visit in 2010 we again lined up the engines side by side on the "yard" trackage in Traverse City, Michigan.

In 2006 I graduated from Michigan Tech with a bachelor's degree in general math and minor in German. I got a job at the Strasburg Rail Road in Pennsylvania. Between October 2011 and April 2012 I laid a 485' loop and I now have the 2-4-0, 2-8-2 and 4-6-0 at my home.

In July 2010 I visited Traverse City for a week. I helped Dad do a boiler wash on the 4-4-2, do some track work and then we brought out our three engines again for pictures since we didn't know when the next opportunity to have them together would occur. (By August 2013 all three engines came to Pennsylvania.)

Our Engines

The First Engine: The Shay

As I have mentioned before, Dad bought a small two-cylinder Shay in 1989 (I was five years old) when he first got into the hobby. The Shay had been built by Dutch Boysen of Yakima, Washington, and he had modeled the train after one that ran near Yakima. He modified the design of a 1.5" scale narrow gauge T-boiler Shay and regauged the trucks to 7.5" gauge. The engine weighed around 100–120 pounds. Dutch had built a ride-on

Figure 225. The Shay is shown on its stand in Dad's shop in 1996. The back of the 0-4-0 (the cab has not been made yet) is visible on the left side of the picture.

car made out of PVC pipes to carry water and I was told he was going to disguise them as logs. He had built two other engines prior to the Shay but his health (dementia) prevented him from continuing in the hobby and his engines were offered up for sale. Dad wanted a small, flexible engine to take sharp curves (the proposed loop in the backyard would have had down to 15–20' radius curves) and since he did not have any machinery (yet) or machining skills (yet) he wanted to buy an engine that was already machined. The sand dome, steam dome and several other small pieces arrived in the mail or UPS and the engine was delivered in a crate. The first summer she lived out in the garage on a stand that Dad made. When we ran her that summer, she was blocked up on some wood on the stand and the flexible drive to the rear truck was disconnected and the cylinders just powered the lead truck as that is where the axle pump was located.

We had trouble running her and had a hard time keeping up pressure and water. The blower was partly clogged and when that was cleaned it was easier to keep up pressure. However the axle pump had a hard time keeping up with the water demand. Dad made a new larger pump that did a good job when the check valves didn't get dirt in them. The hand pump check valves liked to stick so we had to remove the pump multiple times to clean them. We tried installing new checks, but that was difficult as the pump sat in the narrow water leg of the water tank. After messing around with a feedwater heater (15' of ¼" ID tubing in the smokebox) that didn't work, we combined the delivery lines for the hand pump and axle pump and installed a small English/Coles' injector. The injector was finicky and worked only part of the time, but when it did, I preferred to use it to the axle pump. The timing on the engine was off and we had a hard time retiming her as when we got even valve movement and port openings, she would not run and we had to adjust the valve rod to get the engine to run.

While we did get her to run well at times, it was a fight to keep her running well. We would take her down to Bill Fitt's track and run her there. Since the Shay was a slow engine, Dad didn't take her out much during the Tri-State meets as it would hold up the larger rod engines. Several times a rod engine had to push her back to the yard and steaming bays.

As we learned more about steam engines, the more we learned was wrong with the Shay. When the eccentrics were set to the positions they were when she arrived (and per the plans we got with the Shay), they were not set to the proper angles so the valve timing was off. The forward and reverse eccentrics were also pinned together so they could not be adjusted independently of each other. The hydrostatic lubricator was located before the throttle and while the throttle got a decent amount of oil, the valves and the cylinders were starved for oil. The cylinder cocks, with the way they were designed, only allowed water/condensation from the lower half of the cylinder to be drained and not both ends. When the axle pump pumped water into the boiler, the front truck would kick due to the extra force. The gears between the drive line and the gears on the wheels were either not properly lined up or worn/not the correct size as the gears would not drive all the time. With the Briggs boiler there was no way we could clean the coil in the firebox and it was unknown what condition it or the connects for it were in. For whatever reason (possible due to boiler design, pot burner, drafting) the engine did not want to produce steam easily or fast enough when pulling three small logging cars even at a slow speed. (I had tried to convert her over to coal but that failed as I didn't have enough air space between the grate bars ... as I later found out.)

Later on I was told by people who knew Dutch that he finished the Shay in 1982 and around 1983 he could no longer work on his steamers and that the Shay sat around and never had any of the bugs worked out. His other two engines were good runners and steamers; he just never got the chance to fine-tune the Shay.

After we got the 0-4-0 running, we rarely ran the Shay. Around 1997–1998 we traded the Shay to a machinist and welder who machined the larger cylinder castings and made the boiler for the 2-8-2, which was the second engine we built.

#596

Built from a Railroad Supply 0-4-0 kit
Started in 1991, under steam May 1996, finished June 1997
Added lead truck and new tender in 2002
Fuel: coal
Boiler pressure: 130 psi
Valve bore × stroke: 0.750" × .750"
Lead: 0.020"
Bore × stroke: 1.625" × 3"
Driver diameter: 6.000"
Weight: approx. 290 pounds
Tractive effort: about 140 pounds
Blast nozzle diameter: 0.250"
Four notches in Johnson bar for forward.

Figure 226. The 2-4-0 was originally built as an 0-4-0 and first ran in May 1996.

Max load: 9 cars up a 2.5 percent grade, 9 cars up a 4 percent grade really working hard, 17 cars on level track (so far)

The 0-4-0 was the first engine that Dad and I built. She first ran in May 1996 (so her number is 596) and she ran at the last meet at the late Bill Fitt's track.

A nice photograph (fig. 93) was taken at the White Creek Railroad (in Hidden Valley Yard) in the summer of 1996 and my grandfather, Howard Frost, is running her (at 87 years old) and I am riding behind him. It is titled "Ageless Pursuit."

She was the first coal fired engine (the little two-cylinder Shay burned oil in a pot burner) we fired and ran. During the first year or so, we thought we were doing well if we were able to come back after a ½–¾ mile run with 60 psi (it operates at 125 psi) and if we only had to stop once to build up pressure and clean the tubes about every two hours. Compared to the Shay, that was great. Getting a better grade of coal did help.

One of the check valve balls in the axle pump was too small and would like to get stuck and not allow any water to flow to the axle pump. When using the axle pump, the pressure would drop relatively quickly and around 1998 we installed a SuperScale Chicago injector. We loved it and disabled the axle pump. The pressure did not drop as fast with using the injector and it was easier to add water to the boiler. Later we installed a second injector, also from SuperScale.

We also went back over the valve gear in the first couple of years and fixed things that were not machined correctly. This improved the valve events and allowed her to operate more efficiently.

The first lubricator we had used a roll bearing clutch drive, which wore through the mild steel driving shaft in the first couple of years. We then got a couple of Coles lubricators (the first one we made the body, the second was complete) and have used a single pump Coles lubricator on her since.

We started with mechanical cylinder cocks that were solidly mounted in the cylinder casting. After a derailment tore a cylinder cock out of the casting (linkage was damaged, part of the casting cracked off, cylinder cock body was fine) I installed steam operated cylinder cocks that had a sheer "pin/tube" that connected to the casting so the tube would break and not the casting. I also installed a shield under the cylinder cocks to help protect them and made the front foot board strong enough to support the weight of the engine in a derailment. This has come in handy several times.

In the late 1990s we added some washout plug holes to the boiler, bought a small power washer and made up wands to clean the boiler. We also started using water treatment around that time. This helped the boiler to steam better.

Around 2001 we added steel tires to the driving wheels. We had done this to our 2-8-2 previously and found that she had better traction. When we applied steel tires to the 2-4-0, she had a much better footing. The tires have barely worn since then.

Around 2001, after a derailment, I designed and fabricated the lead truck and made her into a 2-4-0. Around that same time we also made a larger tender for her.

Around 2002 or 2003 we made new grates that allowed more air to enter into the fire and this helped the boiler steam better. We then added a stainless steel arch in the firebox and that helped the fire get a more complete burn and not soot up the tubes nearly as quickly. This was the fourth time we had tried using an arch and the first time we got it to work properly.

Around 2005 I designed an equalized brake rigging for the engine as I was having a hard time getting the brake rods adjusted correctly so the brakes applied evenly.

For running when the weather was really dry, I made a screen out of window screen mesh that was tubular in shape and folded over the top. This would fit around the outside of the stack and would clamp on with a hose clamp.

Between 2000 and 2006 we started making some of the valve gear parts, the side rods and main rods out of steel.

***Above:* Figure 227. In the summer of 2002, the 2-4-0 pulled nine cars on the White Creek Railroad. *Below:* Figure 228. Don Frost is coming down the grade, pulling six cars, in the backyard on our track in Pennsylvania.**

Figure 229. The 2-4-0 is in the cut in the front yard of our track in Pennsylvania.

She is now a great running engine, but it took a while (improvements and learning) to get there. The longest train we pulled was nine cars at the White Creek Railroad and she steamed and ran great.

On my own track, she can pull six empty cars up the 3 percent grade and has pulled two cars with three adults and two kids up the grade as well as three cars with three adults.

#1555

Started December 1997, under steam May 1999, finished June 1999 (see fig. 3)
Fuel: coal
Boiler pressure: 130 psi
Valve bore × stroke: 1.250" × 1.000"
Lead: .030"
Bore × stroke: 2.375" × 4"
Driver diameter: 8.375"
Weight: approx. 900 pounds
Tractive effort: about 290 pounds

Figure 230. In the summer of 2002 the 2-8-2 pulled a 31-car train at the White Creek Railroad.

Blast nozzle diameter: 0.375"
Seven notches in Johnson bar for forward
Max load: 31 cars up a 2.5 percent grade, 22 cars up a 3.5 percent grade really working hard, 10 cars with 11 people up a 2 percent+ grade with an S-curve

Dad and I started building the 2-8-2 in late 1996 and she was running late May 1999. She was finished when I was 15 and Dad was 55 so her number is 1555. Dad wanted a large engine to run, but didn't want a common wheel arrangement such as a 4-6-2 or 4-8-4. For her first fire-up she sounded like a jet engine because the piston rings that Dad made didn't seal. He called his father (my grandfather) who in his first sentence listed at least five things we didn't do correctly when making the rings. (He worked at Hastings Piston Ring Manufacturing in Hastings, Michigan, for more than 30 years.) As soon as we installed the professionally made rings there was no more blow by. During her first few runs at the house, one of the main rods (an aluminum casting) broke. I then made two new main rods out of steel and have not had a problem with them.

Her maiden run was far from perfect. The lead truck frame was not perfectly straight, so it wanted to hug the left rail. That was corrected. Later on, the swing hangers on the lead truck started binding up and the truck did not want to swing. It was suggested that we spring the lead and trailing truck separately from the driving wheels to get more weight on the driving wheels. So we modified the lead truck and trailing truck and while it took some tweaking, we got the right combination of weight on the wheels. Not having the lead truck tied into the drivers also came in handy in a derailment where the lead truck frame was bent, as we could remove the lead truck and run the engine back (taking it easy) as an 0-8-2.

When Dad bought the castings, Henry Todaro had purchased the west coast division of Railroad Supply and the casting were offered at a discounted price. We found out why when one of the cast aluminum frame castings going from the main frame to the rear cradle broke. We dropped out the rear driver and I flame cut two new pieces out of steel and installed them in place of the aluminum castings. Around 2005 I also replaced the cast aluminum rear cradle with one fabricated out of steel.

When we made the side rods, we somehow managed to get the centerlines between the right intermediate and right main wheels about 1⁄32" too long. She was stiff, and while running at the White Creek Railroad, pulled out the right main crank pin from the driver center. This happened at the bottom of the hill of the track that parallels the line running by the collection ponds, so we had to push her and the five-car train back up the hill. We made a new side rod that night and the problem was fixed.

Over the summer of 1999 we decided to add steel tires, as a couple of bolt heads stuck out too far and had started chewing away at the back side of the flanges. As it turned out, the driving boxes were accidentally machined to the 7.25" gauge width which gave too much lateral motion.

We were still having problems with the lead truck and lead drivers derailing. We figured this was because we had modified the centering device in 1999 (as was strongly suggested to us by several people who had 20+ years' experience in the hobby) and this modified device (a center roller than pulled on extension springs) was not strong enough. In the spring of 2014 I made a new swing bolster and adjusted the equalizing and spring hangers on the engine. The engine now rides a lot better, stays on the rails and the lead truck now leads the engine into the curves.

Top: Figure 231. The 2-8-2 pauses as it descends the grade on our track in Pennsylvania. *Bottom:* Figure 232. In May 2016 we took the 2-8-2 to the Pisgah Central, a private 7.5" gauge railroad west of Harrisburg, Pennsylvania. The engine ran well. Don Frost is at the throttle.

#1908

Figure 233. Dad and I built the 4-6-0 between September 2000 and May 2001.

Built from an Allen Models 4-6-0 kit and a custom-designed Walschaert valve gear
Started September 2000 and finished May 2001
Fuel: coal
Boiler pressure: 130 psi
Valve bore × stroke: D-valve with 0.750" stroke
Lead: .020"
Bore × stroke: 2.000" × 3.25"
Driver diameter: 8.375"
Weight: approx. 350 pounds
Tractive effort: about 170 pounds
Blast nozzle diameter: 0.312
Six notches in Johnson bar for forward
Max load: 13 cars up a 2.5 percent grade, 10 cars up a 3.5 percent grade really working hard at White Creek Railroad
Max load on my track: abt. 1,900 pounds

Dad and I built the 4-6-0 between September 2000 and May 2001. She was built in memory of my grandfather, Howard Alton Frost, who love steam engines, the color red and was born in 1908. It was decided that instead of using the Stephenson valve gear that we would use a Walschaert valve gear. I referenced one of Dad's old (1920s era) valve gear books and designed and fabricated the Walschaert valve gear for her. We also decided that we would use two super scale injectors and did not install or build the axle pump.

Her first runs in May 2001 on the track Dad had (75' of back and forth) went well so we took her down to the White Creek Railroad in early June 2001. The lead truck derailed easily so we added some more weight on the lead truck and she ran great. After the first year of running we installed steel tires on her drivers (we had installed the tires on the 2-8-2 and were installing tires on the 2-4-0 at this time) and increased her driver diameter from 8" to 8⅜". The original rods were made from aluminum and between 2002 to 2006 the rods were changed over to steel.

In 2002 we decided to make her so she could burn propane and ordered burners

Above: Figure 234. Dad is at the throttle of the 4-6-0 running on the short track in Traverse City around the spring of 2005. *Below:* Figure 235. I designed and fabricated the Walschaert valve gear for her during my junior year of high school.

from LocoParts. Once we figured out how to install the burners and propane lines, changing over from coal to propane takes around 45–60 minutes. In 2006 I rebuilt the brake rigging so that the brakes are equalized per the prototype.

In the fall of 2012 it became apparent that the press fit holding the right main crank pin had failed (measured at a line-in-line fit) so I made a new crank pin and pressed it in. In the fall of 2013 I shortened up the union link to create a more even steam distribution when she is hooked up. She has six notches in forward. In the corner the cut-off is about 82 percent and valve opening is ¼+" and in the sixth notch the cut-off is 50 percent with about .040" port opening.

On my track she has pulled five cars with five adults and her calculated/rated capacity for pulling up a 3 percent grade around a 52' radius curve is 1,880 pounds, so with running into the hill around five mph she was able to pull more than her "rated" capacity. If she tried to do it from a dead stop on the hill, she would not make it.

Rebuilding the Spirit of Traverse City

Figure 236. The Spirit of Traverse City, #400, is a 15" gauge 4-4-2 that ran around Clinch Park and the zoo from 1982 to 2011.

The Spirit of Traverse City is a 3" scale 4-4-2 that ran around Clinch Park (formerly the Clinch Park Zoo) in Traverse City, Michigan, from the early 1980s until 2011. She was built by Lawrence Witherill of Boston, Massachusetts. We believe she was built between the late 1940s and mid 1960s. Lawrence made the plans available for other people to build a 15" gauge 4-4-2. I know of one other 4-4-2 that is stored in the Adirondack Mountains in New York.

After Lawrence passed away, engine #400 was sold to Seymour Johnson of California and we believe he converted the engine to burn oil. In the 1970s the engine was sold to

William Bay of northern Michigan and she sat in storage until the City of Traverse City agreed to purchase the engine and build the track around Clinch Park Zoo. William Bay had the engine completely rebuilt before the city took ownership of the engine. During the winter of 1988–89, Clifford Lenton, an employee of the City of Traverse City, led a complete overhaul of the engine. It is also believed that he built a new door for the firebox that had an air damper to allow some air to be admitted to the fire like the prototype. Clifford passed away in the early 1990s, but his name plaque is still on the back support of the tender seat.

In early June 2004, Dad (Don Frost) and I were asked to re-time the engine as the people who were working on her did not know how to do it. During our initial inspection we noticed that she did not have a mechanical lubricator to lubricate the cylinders. They just had a zerk fitting in the top of the steam chest and they pumped 30W oil into the steam chest once per trip around the 0.45-mile loop. For the timing, we noticed that the link supports had been installed upside down and then we had to reset the eccentrics and lengthen out the eccentric rod on each side by about ⅜". Once we did all of that, she was decently timed and the operators were surprised that they could now operate the engine when she was not in full forward (previously she would only operate in full forward or full reverse before).

In late June or early July 2004 Dad and I were hired to run and take care of the Spirit of Traverse City. We then installed a mechanical lubricator (see fig. 154) purchased from American Model Engineering Supply and in the supply line to the steam chest we installed a check valve and a tee that uses a 5⁄16 MTP angle check valve for a snifter valve.

Figure 237. When we installed the lubricator lines (into the top of the steam chest), we installed a pipe tee and fed the oil in one side; in the other side we installed a 5⁄16" MTP (model taper pipe) angle check valve to be a snifter valve/drifting valve.

With having proper lubrication, the engine rolled much easier. The main problem we have had has been dirt or other foreign matter getting into the oil and keeping the check valves (in the pump and in the line) from working properly.

The boiler is designed, rated and passes inspection for 150 psig but she was only operated at 120 psig maximum as it was believed that this would help save the boiler. The engine pulls three cars (modeled after sugar cane cars) that are around 15 feet long and 1,500 pounds each up a ruling grade of 2.5 percent, which was on a 100–120' radius curve. With the extra drag of not having a lubricator for the cylinders and the reduced boiler pressure, she might be able to just barely get the train up the hill. They would generally run around 20–25 trips a day (six hours). After we added the lubricator and operated her at her designed boiler pressure, we were able to make it up the hill easily and would run an average of 40–50 trips per day (six hours). The old notes from Clifford Lenton showed that the engine regularly ran 40+ trips a day in the late 1980s and early 1990s.

When we started, we were having trouble with having the Penberthy ⅜ NPT injectors working reliably. Both injectors were mounted above the running boards, so they would both be acting as lifting injectors, and were only about ¾" away from the firebox. We cleaned them by taking them apart and soaking them in vinegar. We also noticed that steam and hot water would leak out of the delivery pipe when it was not in use, so we replaced the check valves and the steam valves. This helped quite a bit, but they still had a tendency to overheat so we decided to lower the fireman's side injector below the running board to operate as a non-lifting injector (see fig. 181; just in front of the rear cab support is the glass-bulb fuel filter that we later added). This new location was also farther away from the firebox and it worked well, so during the 2004–05 winter, we also changed the location of the engineer's injector to a lower location.

We also noticed that the lead truck would tend to derail easily. Upon inspection we found that the lead truck was just sitting under the frame but was not bearing any weight. The centering device on the truck frame was sitting into the female counterpart on the frame, keeping the truck from wandering too far, but no weight was being transferred to the truck. We pulled out the truck, added a piece to the centering device on the truck (using a thrust washer as a bearing surface) and re-installed the truck. It took a couple of tries to get the truck to carry the proper amount of weight as to not have too much weight taken off the drivers, but after that, it has been working well.

As we were becoming familiar with the engine, we came up with a list of parts to check each day. Most of this we found out because different parts would loosen up and cause a problem. We also started washing the boiler every month to help keep it clean.

Towards the end of the 2004 season the flues started leaking at the firebox end. We had Roger Schwartz of Schwartz Boiler Shop (in Cheboygan, Michigan) re-roll the tubes. He did not know how long the re-rolling would last, as with leaking tubes, the surface between the flues and sheets is no longer clean. The tubes lasted the rest of the season. During this time we also noticed that a driving wheel was loose on its axle and that the side rod bushings were really loose. We replaced the side rod bushings but then noticed that the front fireman crank pin was chewing through its new bushing. We would not find out why until winter.

Winter 2004–05

The engine and cars were kept in the city's shop during the fall and winter. We tore the engine down to give her a thorough inspection. We found that one of her side rods was bent (it was straightened out easily in a hydraulic press), that the flanges on the drivers were getting thin and that the front fireman's driver was at least 0.100" smaller in diameter than the other three drivers. This is why that crank pin was chewing through bushings. The drivers were also not square on the axles nor were they in gauge.

Top: **Figure 238. Engine #400 is in the City of Traverse City maintenance shop.** ***Bottom:*** **Figure 239. The main and side rods were removed and the valve gear was taken apart in preparation for a complete overhaul.**

We dropped out her drivers and had Grand Traverse Machine turn a new profile on the drivers, bore out the axle holes and made new case-hardened axles. The engine had cast iron driving boxes and solid bearings. During the inspection, we found that the boxes were severely worn and we decided to make new ones and also use roller bearings on the axles and crank pins. The driving boxes did not wear on the frame; the frame had shoes on the front and rear pedestals (closer to the prototype) and we shimmed the shoes to get the axles in tram and machined the side rods in tram. We also made new hardened crank pins, new valves, rebuilt the valve gear, brake rigging and trued up the front truck.

I also made up new cylinder cocks for the engine that were steam operated, so that way if any water built up in the cylinders, or there was water carry-over from the boiler, that the water could force the cylinder cocks open and not damage the cylinders.

Figure 240. I made new steam-operated cylinder cocks for #400. These are an enlarged version of what I made for the 7.5" gauge engines.

We also ended up replacing the throttle. She was set up with a balanced steam dome throttle, per prototype, and no matter how much we lapped the valve into its seat during the running season, we were not able to get the valve to seal. We took the throttle casting out during the winter and sealed it off and tested it under air pressure and soapy water to find the leaks. The throttle valve was sealed in the body; however, the body (cast iron) was full of pinholes. One of the workers tried to braze the thin casting, but after hours of chasing the pinhole leaks, we decided to replace the whole set-up. The new throttle was a 1" NPT steam rated ball valve located in the steam dome.

Summer 2005

We test ran the engine in the middle of May 2005 and she ran a lot better. With all of the drivers being the same size and using roller bearings, she rolled a lot easier. However, the tubes started leaking again in the firebox end and she ran for the first couple of weeks then had to be taken out of service. We took the engine back to the shop. There we removed the trailing truck, lead truck, oil pan, atomizer, cab floor, blast nozzle and took off any valuable gauges or parts that could be damaged easily. We then took her to Schwartz Boiler Shop to have the boiler re-tubed. With the blast nozzle, we noticed that it was partially plugged with furnace cement. We cleaned it, but once we ran her again we found out that the blast nozzle was too large and did not create a good draft. Furnace cement was used to close it up and create a stronger draft.

During this time I designed a new atomizer and oil pan for the engine. The old atomizer did not work very well. It produced a red flame and dripped diesel fuel. The oil pan did not let much air into the fire either. I designed a twin atomizer (modified from Chet Peterson's burner) that produces a partial vacuum at the oil port and produces a bright yellow flame that fills the firebox. When we tested the old atomizer at our shop, we were standing about five feet behind it and it sent out a flame about 2' long and maybe 12" wide. With the new burners, the flames were about 3–4' long and spread over 18–24" and we were impressed with how much more heat we had with standing 5' behind the new burner compared to the old burner. I then designed the new oil pan to give as much firebox volume as possible, use commercially available firebrick as a liner and to allow more air into the fire. I later needed to modify it to allow even more air flow and added a damper to the front around the burner. Just after the tubes were cleaned we took a photo (fig. 86). The rust-colored bar is left over from when we tried using an arch and was the arch support.

This engine was our first experience with an oil burning engine. All of the engines we had built burned coal and one, the ten-wheeler, can be changed over to propane when the weather is dry. We had tried to fire our 2-8-2 on oil, but the atomizers didn't work very well and we didn't like the smell. We were lucky that the burners worked as well as they did because it they didn't, we wouldn't know what to do to get it to work well. We have also had our share of experiences with the fire going out and relighting with a loud boom. People who have heard it said it sounded like a cannon going off and are often surprised to see us just sitting on the tender and barely even flinching even if flames come out of the stack. As we got used to the atomizer, the blowing out and relighting was minimized.

Once the engine came back from Schwartz Boiler Shop, we installed the new atomizer and oil pan. This also greatly improved the steaming qualities of the engine. With the old burner we could make it twice around the track (0.9 miles) but then needed to gain pressure and water. With the new burners and oil pan, I was able to go five times around the track (2.25 miles) with a full load of passengers before I needed to fill up the mechanical lubricator. The pressure and water would be down slightly (140 psi, one-quarter to one-third of a glass of water) but would rebound in a minute or two.

The engine ran well until early August. She then suffered a few premature failures of some of her roller bearings. First, the roller bearing on the large end of the engineer's main rod failed. The engine groaned to a stop and it took a while to work the rod off the pin. We replaced it with a solid bearing and the engine ran for a few more weeks. Then the axle bearing for the engineer's rear driving wheel failed and chewed up the axle and the journal box. We took the engine back to the shop, dropped the axle, pressed the wheels off and replaced the axle (we had a spare) journal box and bearing. After talking with a few bearing specialists, it was decided that even though the bearings were sealed in the journal box, we should grease them (or oil, we ended up trying both and they both worked well) on a regular basis to keep fresh lubrication and to force out any foreign material that may have worked its way past the seal. We have not had a problem with them since.

For the side rods, we changed the front end of the rods back to the solid bearings with about 0.005"–0.009" clearance. When everything was on roller bearings, there was not enough slop. As a wheel drops into a dip or goes up a rise, the axle centerline (and crank pin centerline) lengthens. Roller bearings cannot take changes like that and are not meant for that, so we use solid bearings on the front crank pins to allow the proper slop and not stress the roller bearings.

Winter 2005–06

During the winter we took out the oil pan and added larger air holes and dampers and cleaned the burners. I also took out the original blast nozzle and fabricated a new blast nozzle and blower. I used an NPT plug and drilled it out and made several different sizes so we could change them out and see which one worked the best. We ended up with using a 0.75" diameter blast nozzle (see fig. 91).

Since the engine was running well, we then turned our attentions to the cars and trucks. The truck frames were so worn that you could hold one side frame steady and shift the other side frame forward and back about ¾". To compensate for this, the trucks had double-row, self-aligning ball bearings. The bearings were so wide that when the bearing changed its alignment, the ball bearings would rub on the ends of the journal, wearing the aluminum journal boxes and caps and contaminating the bearings. We had replaced several of these bearings due to failure with a single-row, double-sealed ball bearings, but they failed within a couple months. The axles could also shift over ⅝" laterally in the frames and bearings.

We took the trucks apart and did our best (they were tired) to put them in tram and make them pretty solid. We added spacers to the bolsters where they rubbed on the truck frame to take up extra slop.

Figure 241. This is one of the 15" gauge trucks that went under one of the cars that #400 pulled.

Braces had been added before our time on the train to keep the journals from spreading apart; this is not typical as the bolster is supposed to keep the frames from spreading apart at the journals. We also made spacers to put between the wheels and the bearings to keep the axles from shifting laterally in the frames. We allowed the axles to shift laterally about 1⁄16" to make sure they would not be too tight.

Summer 2006

During the summer, the main problems we had were derailments as we were still taking care of kinks in the track and replacing ties. The trucks now were more solid and could not flex as easily going over rough places in the track. We had a few ball bearing failures on the trucks as some of them still were not as solid as they should be.

In June 2006 I saw an ad in *Live Steam* for a company who was making six-chime whistles. The longest chime was 18" long and the shortest was around 12". Each chime is 1" in diameter. I bought the whistle and made an adapter to fit the whistle valve on the engine. The whistle before it was a small three-chime and the longest chime was just less than 12" and the shortest was around 7". It had a shrill tone and we really did not like it. The whistle I bought has a deep throaty sound and people have said it is like a boat whistle. Others who have heard it stopped by and said that the engine now sounds like a real train and not like a toy.

Figure 242. During the 2005 and 2006 running seasons, Dad, my sister, Rebecca, and I worked on the Spirit of Traverse City (courtesy L. Lee Frost).

General Care

During the running season, we wash the boiler every month. We modified a small power washer to use ¼" tubing with different nozzles to get inside the boiler and wash

Figure 243. Dad is running through the work area. This picture was taken in 2005 and the engine has the small three-chime whistle (courtesy L. Lee Frost).

along the tubes, crownsheet, barrel and water legs of the boiler. We use LSB 8000 as a water treatment and it is easy to show people how to use it and so far we have not seen anything really build up in the boiler.

In order to help prevent derailments due to people putting stones, sticks and other items on the track, we attached paintbrushes to the pilot to brush things off of the track. This does take care of most things, but we have had problems when there are stones in the flangeways of crossings and the engine has derailed several times. Cylinder cocks have been broken this way, but the way I have it designed, there is a sacrificial 5⁄16" tube that connects the cylinder cock body to the plug that is in the cylinder casting. This way the casting is saved from damage in most cases.

For the most part, each winter the valve gear is taken apart and pins and bushings that are worn are replaced. We try to keep a supply of bushings and pins in stock so parts can be replaced easily if something breaks during the running season. We take off the side rods and inspect the roller bearings, the solid bearings for the front crank pins, and the crank pins. The hardened crank pins are just polished and have not worn.

The train ran for about 100 days out of the year and for each day the train makes about 45–50 trips. Each trip is twice around the park (it used to be once when the zoo was still open) for a total distance of 0.9 miles. Over the course of the year, the engine and cars travel around 4,000 miles. This means that each year the drivers are turning around 5.2 million revolutions and the truck wheels are turning over 9 million revolutions. For the amount of use that the engine and cars get, they are lasting quite well.

In 2007, the flues started leaking at the firebox end again. By this time, I was working

Figure 244. People would put stones on the track which could cause the engine to derail so paint-brushes were attached to the pilot to sweep stones off the track.

at the Strasburg Rail Road and I asked for their advice about it. When I told them that the tubes had been rolled in without a copper ferrule (steel tube sealing on a steel sheet) and had not been "seal welded" (more like "heat-sink welded") they said that the tubes would leak in about 3–4 years. The advice was to re-roll the tubes and weld the lip of the tube to the flue sheet so the end of the tube wouldn't overheat. Then roll the tubes again to seal the flue in the sheet. (It was also standard C&O practice to seal weld the tubes in their engines in the last decade of steam or so.) The tubes remained sealed through the end of her last season, 2011 (see fig. 137).

By 2008 Dad asked me about the flange thickness and how thin we should let the flanges become before replacing the wheels. I asked at the Strasburg Rail Road and was told the best thing to do would be to scale down the condemning limits that the FRA uses. I drew up the sketches with the condemning limits and all the wheels were still good; a few were getting close, though. I made up drawings for new wheels and axles and emailed them to Dad. For the 2009 season, the worst wheel sets had been replaced and for 2010 all the wheel sets had been replaced.

Figure 245. New wheel sets (wheels and axles) have been machined and installed in the truck frames.

Figure 245 is of the new wheel sets in the frames. Figure 241 shows the trucks with the old wheel sets (the same wheel sets that it had when it first started operation in the early 1980s). The truck frames were still not in the greatest shape and they did not always hold the axles in tram so sometimes one axle set would wear severely due to being out of tram.

Top: Figure 246. After doing a boiler wash in September 2006, Dad and I ran #400 around the track for our own enjoyment. Dad is starting on the hill with the cylinder cocks open. *Bottom:* Figure 247. Engine #400 is stopped in the station. The safety is popping and she is ready to go.

Figure 248. Dad is running up the hill on the east end of the park.

Figures 246–248 were taken in September 2006. At this time the zoo had been closed and it was uncertain if the train was going to continue to run. Dad and I had fired her up after doing a boiler wash and we decided to run her around and take pictures since we didn't know if she was going to run again. She did run again in 2007, but this was the last day that Dad and I ran her together. She ran through the 2011 season and then the City of Traverse City decided to get rid of her in favor of a splash pad.

Figure 249. I am taking the train west through the work area (courtesy L. Lee Frost).

Engine

Built by Lawrence Witherill of Boston, Massachusetts, ca. 1948–1962

Rebuilt by William Bay in 1979

Rebuilt by Clifford Lenton in 1989

Rebuilt by Don and Jeff Frost in 2004. During the last rebuild all the running gear was rebuilt, two new piston valves were made, one new piston was made, new brake pistons and liners were made, new driver axles were made and all drive wheels were re-profiled, a new blast nozzle was made, new steam cones for the injectors were machined, a new atomizer and oil pan were designed and made, new cylinder cocks were made and a new lubricator was added

Boiler retubed in 2005

Fuel: diesel

Boiler pressure: 150 psi

Valve bore × stroke: 2.075" × 2.000"

Bore × stroke: 3.780" × 4.500"

Driver diameter: 15.300"

Boiler capacity: approx. 30 gallons at running level

Weight: approx. 3,000 pounds dry; approx. 3,200 pounds wet

Tractive effort: about 530 pounds

Calculated pulling capabilities: 104 loaded cars (2,500 pounds each loaded) on dead level straight track; 16 cars up a straight 1 percent grade; 8 cars up a straight 2 percent grade; 3–4 cars up a 2 percent grade on a 110' radius curve

Figure 250. View of the backhead and the controls and gauges.

Figure 251. Don Frost is taking a full trainload of passengers up the hill (courtesy L. Lee Frost).

Tender

Fuel capacity: approx. 14 gallons
Water capacity: approx. 80 gallons
Weight: approx. 300 pounds dry; approx. 1,000 pounds wet

Cars

Length: approx. 16'
Width: approx. 32"
Capacity: 10–12 people (six rows of seats that may fit two people each)
Wheel diameter: 8.625"
Wheel material: 4140
Weight: approx. 1,500 pounds empty; approx. 2,500 pounds full with passengers

Getting Back into the Hobby

In late December 2006 I moved to Lancaster, Pennsylvania, to work at the Strasburg Rail Road. I had an apartment so I didn't have any room for live steam. On a couple of the trips back to Michigan Dad and I would run some of the engines we had built. In 2010 I bought a house, as I wanted a decent property to have a live steam track around the house. The property is .7 acres, is reasonably flat and has a detached garage with a smaller section inside that is heated. I created a garden, with about 35' of track running through it.

In August 2011 I bought some panels of track from a friend. Since he had 4.75" and 7.25" gauge equipment the panels were dual gauge. Since I have 7.5" gauge equipment, I had to move one rail ¼" and I decided to build a dual gauge track as there are several friends in this area who have 4.75" gauge equipment.

I started work on the track on October 16, 2011. After measuring the yard I decided to use a 52' radius curve with a half circle in the front yard and a half circle in the backyard with 80' of straight between the two half circles. The 2-8-2 can take down to a 40' radius but I wanted to use a broader radius than the minimum. With the placement of the house and detached garage and the use of a broad radius curve, I couldn't really vary from an oval.

I staked out the right of way and used a tape measure to measure the radius from the center of the curve. Figure 255 is taken in the front yard and the track will cross the driveway here. The corner of the house can be seen in the upper left hand side of figure 256. This is also the front yard where the hump in the ground is. There is about a 64" rise in the ground from the driveway to the side of the house. The ground before the hump rises at a 3 percent grade. I staked the curve in the backyard (fig. 257) and then used a string line from the ends of the two half circles to stake out the straight sections of the right of way.

Figure 255. I plotted the stakes out for the right of way and chose a spot on the driveway that would be the easiest to cross.

Opposite, top: **Figure 252. Rebecca guides a trainload of happy passengers down the hill on the west end of the park (courtesy L. Lee Frost).** ***Middle:*** **Figure 253. Jeff approaches the trestle with the cylinder cocks open. About one and a half train lengths beyond the trestle is the station (courtesy Donald S. Frost).** ***Bottom:*** **Figure 254. Don Frost approaches the top of the grade.**

Top: Figure 256. The front yard is around a 3 percent grade and then has a hump in it by the tree near the front of the house. *Bottom:* Figure 257. The backyard is mostly a gentle slope but does have a dip by the one corner.

On October 17, 2011, I started grading the right of way. Due to the height difference, I decided to use a 3 percent grade in the front yard to minimize the cut through the hump and in the backyard I would do half the grade with a 2.5 percent and the other half with a 3 percent. I wanted to have a removable piece of track across the driveway so I started grading at the drive way.

Top: **Figure 258. I dug through the sod and graded the ground.** ***Bottom:*** **Figure 259. Grading the yard.**

To grade the right of way I used an eight-foot-long 2 × 10, a 48" level and a one-inch tall piece of steel. A 3 percent grade is also called a 1:33 gradient, which means a 1" rise/fall in 33". I would place the board on the grade, the level on the board and the 1" piece of steel under the 33" mark on the downhill side of the level. When the level was level, I had the 3 percent grade.

Figure 260. By the end of the day on the 17th I had put 30 feet of track on the ground and staked it in place until I tamped it in.

By October 26 I had 70 feet of track laid. I did not use stone for ballast as stone is not considered a permeable surface and the amount of stone I would need to use would have required a permit. Since I didn't want to get local politics involved, I used the dirt that I had dug up as "ballast." When tamping the track, I superelevated the outside rail on the curve by ⅛"–³⁄₁₆".

Figure 261. I had to dig through the raise in the yard, grade the ground for the track and then slope the sod to the track. I would lay track before continuing to dig the roadbed. At this point, the track ends in a wall of dirt.

I had 90 feet of track laid on December 17, 2011. At the end of the track (fig. 261) a wall of dirt can be seen where the ground picks up a lot. This is at the deepest part of the cut, about 17" down to the base of the roadbed. The wall of dirt is a few feet beyond the end of the track. It was a lot of work to dig all of the dirt out. I had a good size pile of dirt in the front yard near the detached garage and another large pile behind the house. Digging in this area I hit a hard section of blue colored and very wet clay. It wouldn't surprise me if the grade is slightly steeper there.

Figure 262. Beside the house, I had less digging to do. The trench for the roadbed can be seen. At this point, I didn't slope the sod to the track.

Figure 262 is from January 1, 2012 (120 feet of track) and shows how I dug the cut. I started out by digging a straight walled cut about 28" wide and I would grade the floor of the trench. I then cut the sod on the sides into sections and removed the sod. I then cut back the sides of the trench on an angle and replaced the sod. The cut is around 9–12" deep here.

There is only about 20–30 feet of level track at the top of the grade before it starts back down again. When I laid the track I did not fine tune the grading; that was done later when it was warmer.

By February 5, I was through the garden (which had just been started the previous year) with 240 feet of track on the ground. Just beyond the top of the grade, the ground drops off so I had to do a fill. The track is descending on a 2.5 percent grade and reaches the ground level again near the end of the garden.

Figure 263. By early February 2012 I had 240 feet of track laid (just under half of the length of the loop). The top of the grade is near the end of the deck.

Figure 265 was taken from the back porch on February 5. Just in front of the garden is the large pile of dirt that was removed from the cut in the front yard. (The pile in the front yard was about one-third this size.) Some of the dirt has been used as "ballast" and as fill. All of this was dug by hand and moved one wheelbarrow at a time.

I had 340' of track laid by March 10 (a mild winter helped with laying the track). The backyard required a fill that, at its tallest point, is about 18" tall to the bottom of the tie. Once the loop was finished, I widened out the fill with the leftover dirt.

I then started working on the roadbed from the other side of the driveway, working back to where the end of the track was. I again started off the level of the driveway and leveled off the roadbed within 50 feet of the driveway.

Top: Figure 264. The track goes through Veronica's garden and is heading back downhill. I had to use some of the dirt that I had dug out of the cuts for some small fills. *Bottom:* Figure 265. The track can be seen passing through Veronica's garden in this view from the deck. The big pile of dirt is one of two piles I made from the cut in the front yard.

Top: Figure 266. By mid-March 2012 I had laid 340 feet of track. The dip in the backyard required around 18 inches of fill. *Bottom:* Figure 267. I then shifted gears to finish up the roadbed and started at the driveway and worked my way towards the backyard.

Figure 268. The roadbed was complete on March 12, 2012.

Figure 269. By March 25 I had 420 feet of track laid.

I would assemble the track in 20-foot sections in the garage and then carry them out to the roadbed. I used treated 2 × 2s for the ties. They are around 13⅝" long and I get seven ties from an 8-foot length. I used sheet metal screws that have a washer built in the head to hold the Culp rail to the ties. For the joint bar bolts, I used 6–32 machine screws.

Figure 270. By April 1, I had the track back to the driveway.

By April 1, all that was left to do was to build the removable piece that goes across the driveway. The loop was completed on April 8, 2012, when I built the 14-foot removable piece to go across the driveway; the piece would be installed each time we ran the train. (In 2016 we had the driveway cut and the track was installed in the driveway. This makes it a lot easier to run the train.)

The entire loop is about 485 feet long and was completed in just under six months. The gondola in figure 271 was built for Bill Fitt in the 1970s and is still lettered (although it is worn) for the Meauwataka and Wildwood Railroad. On one of their trips to visit, Mom and Dad brought several cars from Traverse City.

While the loop was complete, I still had to do final alignment and tamping. That took another couple of weeks to do and it gave a really solid track.

To bend the rail for the curves, Dad brought me the rail bender that he had made in mid-1990s. A fellow Tri State Locomotive Club member had let Dad borrow his rail bender and Dad made this one using the borrowed one as a pattern.

Three rollers are placed in a "T" shape with the bottom roller being an adjustable

Top: Figure 271. The loop was completed by making a removable piece that sits on the driveway. *Bottom:* Figure 272. I made the rail bender out of 2 × 4s.

roller to adjust radius of the curve on the rail that is pushed through it. The rollers ride on shoulder bolts. The thickness of the rollers should be thin enough to fit in the web of the rail. An improvement on this set-up would be to use ball bearings instead of home-made rollers and to have the rollers closer together so the length of the unbent/curved section of the rail won't be as long.

Figure 273. This is the rail bender that Dad made copying one that a fellow live steamer made.

Once the loop was completed, I built a switch to head to a steaming bay. I installed the switch at the end of the curve in the backyard and brought track along the side of the garage. I built a small retaining wall as the track was climbing. I built a wooden ramp and steaming bay (which is 10 feet long) as I don't have any machinery for metal yet. The height is close to the height of the rolling stands that the engines are on when they are in storage.

The engines are on their own rolling stands that are easily rolled out from the garage and onto the driveway. The ends of the stands have an adaptor to connect to a bridge that runs between the stand and the steaming bay.

Due to the lineup of the steaming bay, the rolling stand will be just off the driveway. With the 4-6-0 and 2-4-0 (with its tender on the stand) being 350+ pounds, moving them on and off the pavement is not too difficult. The 2-8-2, since she weighs around 900 pounds, is a different story. In figure 275 the boards that are on the ground are set for the wheels on the Mikado's rolling stand so they don't sink into the ground. A driveway extension done in 2016 makes it easier to get the engines in and out.

In June 2012 Dad and Mom brought the 2-4-0 from Traverse City and then in August 2012 they brought the 4-6-0. With the White Creek Railroad being the nearest track (2.5 hours away) Dad wasn't running the engines much so in 2013 they also brought down the 2-8-2. The engines have run more here between 2012 and 2014 than they were run between 2004 and 2012. Since Dad doesn't have any engines in Traverse City, he is now building a 7.5" gauge Climax that will probably be running around 2018.

Top: Figure 274. The stands roll out to the steaming bay (in 2016 we had the driveway extended so the stands are on the asphalt); a ramp is used to go between the stand and steaming bay. *Bottom*: Figure 275. A bridge runs between the stand and the steaming bay.

I keep the engines in the heated room in the detached garage and keep incandescent light bulbs in their fireboxes to keep the boilers warm and to keep condensation from forming. With the bench that was there when we bought the house, it is a tight fit. At this point I don't have much in the way of machinery, but in time my plan is to start acquiring machinery.

Top: **Figure 276. Part of the detached garage has climate control and this is where the engines are kept. The Mikado and 2-4-0 are pulled in straight through the door. The 2-4-0's stand is eight feet long (same length as the Mikado's) and is just long enough for the engine and tender.** ***Bottom:*** **Figure 277. The 4-6-0 and its tender are kept just to the right of the door.**

During the warmer months I usually get an engine out every few weeks. Because the 4-6-0 and 2-4-0 are easier to get out than the 2-8-2, they are generally run more, with the 2-4-0 being run more than the 4-6-0. The loop is about 485 feet long, so 11 trips around is just over one mile. When I am running for my boys, the engines are typically running for about 1.5 hours and run around three miles. While running on a long track is fun, I have enough track work on the 485' loop to keep me busy and not wanting more. Since the loop has been laid, I have been tweaking the track, grades, cross level and the curves and adding water drainage for the track near the garage.

Figure 278. Due to the water runoff, I had to install ways to get the water to flow under the track instead of creating a big puddle and flowing over the track.

There are times when the kids will ride the train for at least an hour and that gives her time to do some gardening. The kids also enjoy sitting in front of me on the tender and wearing my gloves (which are way too large, but they'd rather wear mine than their own) and ringing the bell, tooting the whistle and pulling the throttle.

Introducing People to the Hobby

One of the things I really enjoy is introducing people to the hobby and watching them have fun as they take their first ride behind a steam engine or pulling the throttle for the first time. Building, designing, running and tweaking steam engines can be fulfilling and so can making people happy. In order of this hobby to thrive, the younger generation needs to be introduced to the hobby, enjoy it, be involved as well as being safe and knowledgeable. I enjoy answering questions, passing along knowledge and encouraging people in the hobby.

All of the valves on the engines I have are set up in a similar fashion and they can easily be accessed by someone sitting on the tender or walking beside the engine. One of the reasons why I like a short loop like mine is it makes it easy to walk beside the engine to supervise a new operator, use the injectors, shut off the throttle and apply the brakes. On a long track, I would sit behind them without being able to see much or access much. With my track being 485 feet long, the engines can run a couple of times around the track without needing water added if they start off with a little more than half a glass and can typically run four trips before needing the fire tended. When someone is new at running the engine, I will explain why having water in the boiler and water glass is

Top: Figure 279. Veronica will also occasionally take a turn at the throttle. *Bottom:* Figure 280. Both my boys enjoy riding and are starting to learn how to run the engine (courtesy L. Lee Frost).

Top: Figure 281. Matt takes a turn at the throttle on the 2-4-0 while Mike enjoys his ride. *Bottom:* Figure 282. Chris is approaching the top of the grade while the boys look on.

Top: **Figure 283. Chris guides the Mikado down the hill.** ***Bottom:*** **Figure 284. Eric enjoyed running the ten-wheeler in October 2014.**

Top: Figure 285. In October of 2013 we had a group of people over from church and they rode on the train. Paul is at the throttle of the Mikado. *Bottom:* Figure 286. Paul has just tended the fire while drifting down the hill.

Top: **Figure 287. When my parents visit, Dad gets a chance to run the engines. Here he is coming down the grade in the backyard with my mom, Veronica and the boys riding.** ***Bottom:*** **Figure 288. When I would come home for spring break from Michigan Technological University, I would shovel off the track (there would still be snow on the ground in northern lower Michigan) and we would get an engine out. Here Dad is running the 2-4-0 in the spring of 2005 or 2006.**

important, how to stop, how to apply the brakes, the basic use of the Johnson bar and the throttle, where the top of the grade is and where the steep downhill begins. While I will show them how to use the injector, I don't expect them to use it if this is their first time. I will walk beside the engine (supervising and explaining things at the same time) and use the injector so they can run several times around the track before needing to stop to tend the fire.

Top: Figure 289. Mike has a turn on the Mikado in 2016. *Bottom:* Figure 291. Dad and I built the 4-6-0 in memory of my grandfather, Howard. He was born in 1908 and loved the color red. He died in July 2000 and Dad and I started on the 4-6-0 in September of 2000 and finished her in May 2001. This picture is taken around 2015. The boys are also developing a love for steam locomotives.

In this section are pictures of people having fun in the hobby; some of them have been in the hobby for a while, others are just being introduced. Some of the pictures are 10 or more years old while others are more recent.

Whether it is having friends and family over to run the 7.5" gauge engines, being on the train or engine at Strasburg, operating the 15" gauge steam engine at Strasburg or the 15" gauge train in Traverse City, I have noticed that most people enjoy being around and riding behind a steam engine. There are not too many frowns. Making people happy like that is addicting and I enjoy seeing people having fun.

Opposite, top: Figure 292. The parts of a locomotive: (1) smokebox; (2) boiler; (3) sand dome; (4) steam dome; (5) tender (holds fuel and water); (6) cylinder block; (7) lead truck and lead truck wheel; (8) driving wheels (four per side on this engine, eight total); (9) trailing truck and trailing truck wheel; (10) tender trucks and tender truck wheels; (11) cab. *Bottom:* Figure 293. The parts of the driving gear: (12) piston rod; (13) crosshead; (14) wrist pin; (15) main rod; (16) side rod—main; (17) secondary side rod; (18) knuckle pin.

Appendix A: Major Parts of a Steam Locomotive

Figures 292 through 295 show the parts of a steam locomotive, from the engine to the driving gear to the valve gear to the plumbing. Photographs of a 7.5" gauge 2-8-2 Mikado steam locomotive have been used.

Figure 294. The parts of the Walschaert valve gear: (19) lead and lap lever; (20) union link; (21) valve crosshead; (22) radius rod; (23) link (and link block inside); (24) trunion and trunion/link pivot; (25) eccentric rod; (26) eccentric crank and eccentric crank pin.

Figure 295. The basic plumbing parts: (27) water delivery fitting into boiler; (28) shut off valve to boiler (in case the check valve sticks open); (29) check valve (one way valve, allows water into boiler but not out); (30) delivery line; (31) injector (uses steam to force water into boiler); (32) water line from tender; (33) mechanical steam oil lubricator (for cylinders); (34) steam oil delivery lines (to cylinders); (35) safety valves; (36) washout plugs; (37) blowdown valve (drain for boiler).

Appendix B: Common Questions

How does a steam locomotive work?

Steam locomotive boilers are generally horizontal fire tube boilers, meaning that they lie along a horizontal plane rather than being vertical and the tubes inside the boiler carry the heat and gases/smoke from the fire. Generally the firebox is on the rear of the boiler. The firebox consists of two metal shells (one fits inside the other) and water fills the space between the two shells. The fire is in the inside shell and the outside shell is what can be seen from outside the engine. The round part of the boiler is called the barrel and has tubes that run from the front of the firebox to the front of the boiler. The tubes carry the heat and gases/smoke from the fire to the smokebox which is attached to the front of the boiler. The inner shell of the firebox and all the tubes are surrounded by water to keep them from overheating.

The fire in the firebox (fuel can be coal, wood, oil or propane in miniature) heats the inside shell of the firebox. The heat is transferred through the firebox metal to the water and heats up the water. Most of the boiling of the water takes place in the firebox. The heat and gases/smoke pass through the tubes and the heat passes through the tubes to the water in the barrel and acts somewhat like a preheater for the water. The gases enter the smokebox and exit through the smokestack.

The heat from the fire will make the water boil and will produce steam. Steam is collected in the steam dome, the highest part of the boiler which provides the driest steam. A throttle can be in the steam dome or in the smokebox (shown in drawing 47). The throttle will control the steam flow from the boiler to the cylinders. The valve in the cylinder block will admit steam to one side of the piston and exhaust it from the opposite side.

Figures 296–301 show how an inside admission spool valve (so named as it is in the shape of a spool that holds thread. Inside admission means that the steam from the boiler enters the steam chest and surrounds the center of the valve) with direct motion Walschaert valve gear (when running in forwards, the valve movement is the same direction as the movement from the eccentric crank) moves the valve to direct the steam from the boiler into the piston and exhausts the steam from the cylinder for running in forwards. The pressure of the steam being exhausted through the valve and exhaust passage ways is a lot lower than the pressure entering the steam chest; the exhaust pressure could be around 10–15 psi or lower. When the engine is working, the exhaust can be heard but if a light, drifting throttle is used, the exhaust might not be audible (less steam is being used so less steam is being exhausted).

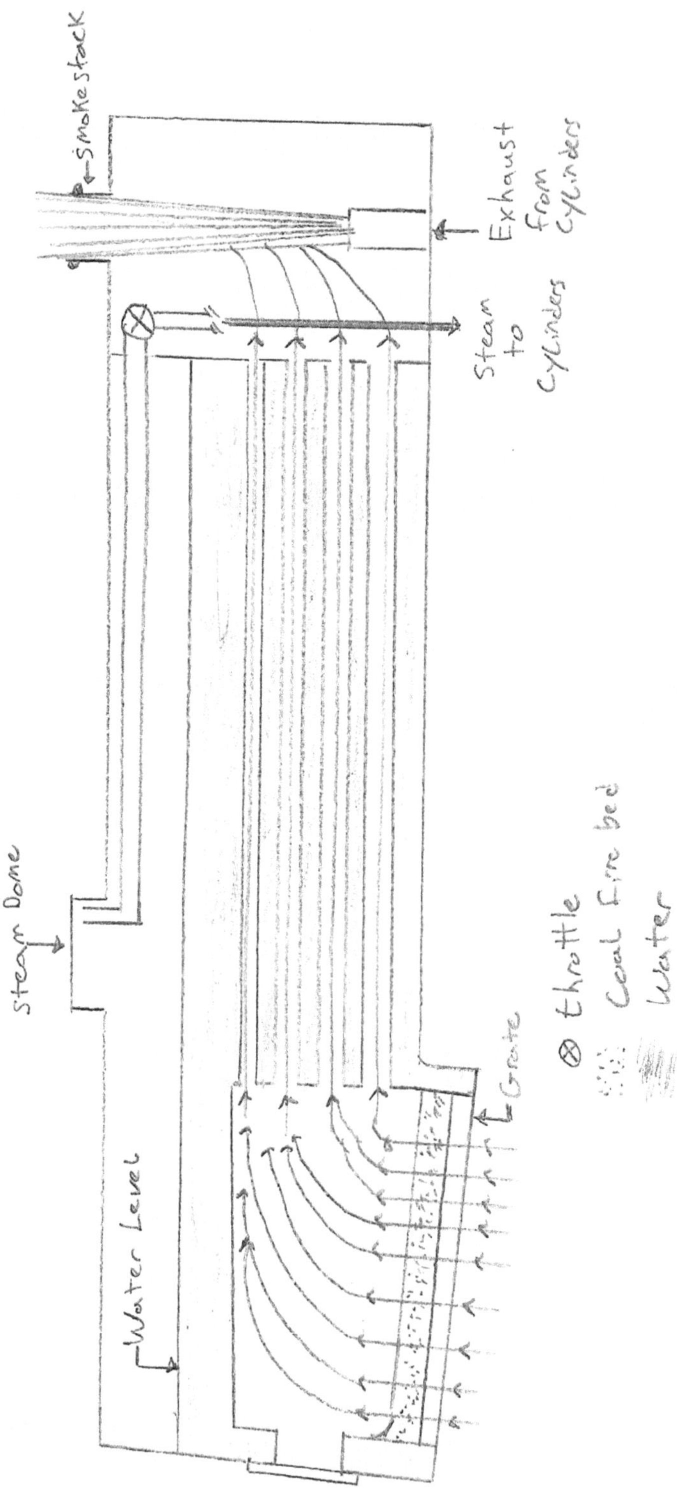

Drawing 47. A cut-away drawing of the boiler shows the water in the boiler, the firebox, fire, tubes and smokebox.

Top and middle: Figures 296 and 297. With the piston being at the front end of its stroke, the valve lets steam in the front of the piston (solid line) to push it back and exhausts the steam from back side of the piston (slashed line). *Bottom:* Figure 298. The valve has stopped admitting steam to the piston but the steam in the piston chamber is still expanding and pushing the piston backwards.

Top and middle: **Figures 299 and 300. With the piston being at the rear end of its stroke, the valve lets steam into the back of the piston (solid line) to push it forward and exhausts the steam from the front side of the piston (slashed line).** ***Bottom:*** **Figure 301. The valve has stopped admitting steam to the piston but the steam in the piston chamber is still expanding and pushing the piston forward.**

To run in reverse, the valve motion is reversed.

Figure 302 shows the valve gear set for running in forwards and the steam from the boiler pushing on the piston (and the steam being exhausted) to rotate the drivers to run forwards.

Figure 303 has the wheel in the same position and the valve gear is set for reverse. The steam is being admitted back side of the piston to pull on the main rod to rotate the wheels to run in reverse.

Top: **Figure 302. When the valve gear is set in forward, the steam is admitted to the front of the piston to push on the main rod and rotate the wheels to move forward.** ***Bottom:*** **Figure 303. When the valve gear is set in reverse, the steam is admitted to the rear side of the piston to pull on the main rod and rotate the wheels to move in reverse.**

The power of the steam acting on the piston is transmitted to outside the cylinder by the piston rod connected to the crosshead (see fig. 293). The main rod is connected to the crosshead by the wrist pin on the front and is connected to the main crank pin on the main driving wheel. The main rod transmits the force from the crosshead to the main driving wheel. The side rods connect all the drive wheels together and transmit the power from the main driving wheels to the non-main driving wheels. To give the side rods flexibility as the wheels move up and down over the track, a joint is added called a knuckle pin. It transmits the force while allowing the rods to have vertical flexibility. Rod engines that have more than two connected driving axles will use knuckle pins. Engines with three driving axles will have one knuckle pin per side, engines with four driving axles will have two knuckle pins per side and engines with five driving axles will have three knuckle pins per side.

What is the chuffing noise that is heard when a steam engine is running? What does it do?

The chugging or chuffing noise that is heard when a steam engine is running is the steam being exhausted from the cylinders. When the valve exhausts the steam from the cylinders, the steam will travel through passageways, up the blast nozzle and up the stack. (See drawing 47.) The exhaust steam creates a venturi effect as it passes through the blast nozzle and creates a partial vacuum in the smokebox. With the smokebox being sealed, the only way for air to enter the smokebox is up through the firebox, the fire and through the tubes. This brings in more air for the fire and is called the draft. In a coal-fired engine, this increase in the air being pulled through the coalbed makes the coal burn hotter (sort of like blowing on a wood fire makes it burn hotter). In this way, a coal fired engine is a bit "self-regulating" because when it is working harder and needs more steam, the increased draft will increase the temperature of the fire and help increase the steam output. The same is not so true for oil or propane burning engines.

How do the wheels stay on the track, or follow the track, and how are they attached to the axles?

The inside face of the wheel has a rim that is a larger diameter than the tread (the diameter that rides on the rail). This rim is called the flange and it is on the inside of the railhead. Both wheels on the axle (aside from blind drivers [flangeless drives] on a steam locomotive) have flanges that are inside the rails. The axle set can only move so far to one side before the flange will bump up against the rail and keep the axle set from moving any more to the side. The tread is tapered (1:20 taper, 2.86 degrees) to help the wheel set find its happy, centered home on straight track (this keeps the wheel sets from wandering too much and causing the flanges to bump into the rail). This taper also helps on a curve because the axle set will drift to the outside of the curve so the outside wheel will ride on a slightly larger diameter than the wheel on the inside of the curve. The fillet is a radius transition between the tread and the flange and helps keep the flange from contacting the railhead. The flange has a radius on its edge to keep it from digging into the railhead (a sharp flange will wear the rail and will dig in and is more prone to climbing the rail and derailing). The radius between the tread and outside face will minimize damage/cutting to frogs (switches, crossovers) and switch points as well as leave a relief so when the metal starts to roll over it doesn't cause damage.

Drawing 48 shows the bearing/journal surface on the axle being on the outside of the wheel as is common with car axles and diesel axles. Since this book is written for the

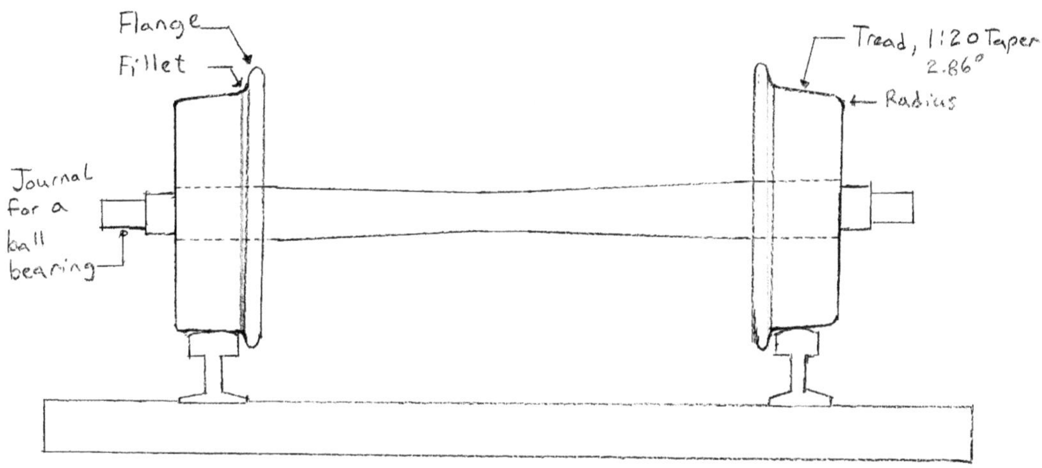

Drawing 48. The wheel set has the parts of the wheel called out and show how the flanges are on the inside of the rail.

live steam hobby, I labeled the journal to use a ball bearing (prototype would be plain bearings or roller bearings) as ball bearings are now commonly used in the hobby for axles.

The wheels are pressed onto the axle (interference fit) and in the smaller scales they can also be loctited to the axles. For a press fit, the wheel seat on the axle is slightly bigger than the bore in the wheel. A general rule of thumb is for under 1" diameter, the axle (or shaft) would be 1.002–1.003 times the bore diameter. For over 1" diameter the axle (or shaft) would be 1.001–1.002 times bigger than the bore diameter. The press would hold the wheel on the axle and would do the driving/powering. Keyways in the wheel/axle on the prototype are for locating but in miniature have also transmitted the power between wheel and axle. The end of the bore or end of the axle/shaft should have a small taper to help the parts slide together without galling. The parts should also be lubricated to prevent galling as well. The axle/shaft does not need to have a shoulder

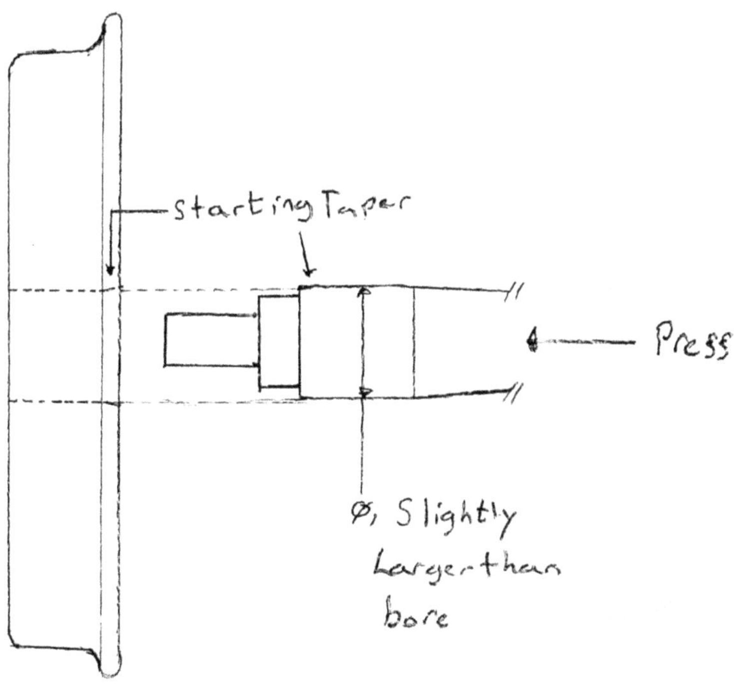

Drawing 49. This shows how the axle is pressed into the wheel.

turned on it to know when to stop the press; if measured carefully when pressing, the wheel can be pressed on square and accurate to its final position.

How are the wheel sets held in place?

The bearings for the driving wheel sets of a steam locomotive are mounted in driving boxes. The boxes are in the steam locomotive frame and can move vertically but are held in place otherwise by the pedestals on the front and back of the driving box.

Lead and trailing trucks have journal boxes that can be mounted to the axle either inside or outside the wheels. The 2-8-2 has an inside frame for its lead truck (fig. 305) and an outside frame for the trailing truck (fig. 306). With both trucks, the journal boxes can move vertically but are otherwise held in place by the pedestals (on the front and back of the boxes) that are attached to the truck frame.

Diesels and cars use frames to hold the wheel sets in their appropriate locations. For freight cars (and a lot of tenders) the journal boxes are rigidly mounted to the truck frame. Figure 307 is of the 2-4-0's tender truck. The journal boxes are rigidly held to the frame but the truck is flexible enough to allow one wheel to pick up while the others remain on the track. Diesels and most passenger cars use trucks that have boxes that ride vertically in pedestals mounted to the truck. On some small cars that only use two axles, the car frame will have pedestals mounted on it and the journal boxes will have vertical motion in the pedestals.

Figure 304. The top of the driving box can be seen (under the leaf spring) on the 2-8-2. The box can move vertically but is held in place by the pedestals on the front and back of the driving box.

Left: Figure 305. The left journal box of the 2-8-2 can move vertically but is held in place by the pedestals on the front and back. The pedestals are attached to the lead truck frame. *Right:* Figure 306. The trailing truck journal box is on the outside of the wheel and is mounted in the trailing truck frame. The journal box can move vertically.

Figure 307. The truck frame of the 2-4-0's tender holds the axles at the proper spacing. The truck is flexible to allow one wheel to pick up while the other wheels stay on the rail.

How fast will it go?

My initial response is, "Too fast!" Then I explain that scale steam engines are not go-karts or racecars and that with the weight of the train there is a lot of momentum (speed times weight) and it is hard to stop quickly. If there is a derailment, things can get ugly really quickly if cars turn over or land on someone. The fastest I like to run with an empty train is around 6–7 mph (8.8–10.3 ft/sec) however with passengers I generally run around 5–6 mph (7.3–8.8 ft/sec) at the fastest but I normally run around 4–5 mph (5.9–7.3 ft/sec). I have had a light engine (1:8 scale) up to 10 mph (14.6 ft/sec) once and that is flying. Probably the 4-6-0 could do 20+ mph (29.3 ft/sec) if given a chance. I have heard people say that they had their 1:8 scale engines do 15 mph (22 ft/sec). Once I was told of a story of an operator who had a scale 4-4-0 (12" gauge or thereabouts) and was running it at 20 mph; it derailed on a fill. A kid on the train fell off and the equipment (engine or car) rolled over him and killed him. When people say that 10 mph is a scale 80 mph and passenger engines ran that fast, my response is to say that they also had brakes on all the cars and the track was in very good condition. Pretty much no one has a track built or maintained to scale 80 mph standards; derailments are likely when running too fast on track that cannot handle it.

Do you use computers to assist the operation of the engine?

Someone asked me this of a 15" gauge steam engine. When I said "no" he was a bit surprised and replied, "If the water gets too low, bad things can happen, right? What prevents that from happening?" I said that it is the engineer's responsibility to keep the water at a proper level and to run the train properly. He then said, "But computers make things safer and more reliable." Doing my best not to laugh, I responded by saying that computers are only as good as the information that is given to them and the best type of operator is a person who is skilled and qualified (qualified on the equipment, knowledgeable, attentive, runs the equipment well and understands it well) and that I didn't think a computerized program could do as good of a job as a qualified engineer. No two trips are the same and the engineer needs to quickly adapt to the differences and think on his or her feet. He responded with an "oh" and walked away.

Why can I add more water to the boiler when the fire is hot without dropping pressure?

The circulation of the water in the boiler is driven by the heat from the fire. If the fire is hot (from working up a hill or a strong forced draft from the blower) the circulation will be increased. The sheets in the firebox will be exposed to a higher heat (coal can burn between around 1,900°F to around 2,400°F) which will be transferred to the water. Even after the draft is decreased, the fire will burn hotter for some time. When the water is added to the front of the boiler, the cooler, denser water will descend to the bottom of the boiler and flow to the mud ring. With the hotter fire, more heat transfer through the sheets and faster circulation, the water that is added can be heated up more quickly and the overall temperature (and pressure) of the boiler won't drop nearly as fast as when the fire and sheets are cooler and the circulation is slower. It is possible to have the pressure rise when working an engine with an injector on. The smaller the engine, the harder that is, but it is possible in the 1:8 scale RRSC 0-4-0.

I have heard it said that the best time to add water to a boiler is when the engine is working uphill, as the working of the engine and its movements/vibrations/running down the track increase the circulation. The sloshing of the water by running down the track does not increase circulation but it is the extra heat from the fire when working the engine that increases the circulation.

When adding water, it is good practice to not drop the pressure more than 5–7 psi at any one time. This is to help keep the boiler water temperature even and minimize thermal shock on the boiler.

Will making the center drivers blind make my engine able to take tighter curves?

Whenever I am asked this, my answer is, "Maybe." Taking tighter curves is more than just turning the flange off a driver. Blind tires were generally wider than flanged tires as the blind drivers can move around more on the rail head. However, there are other potential limitations with the engine than just flanged drivers. When heading around a tighter curve, the leading flanges will also take more abuse which can shorten their life. For an engine to take a tighter curve, the lead truck and trailing truck (if any) will need to swing more. There may be interference between the truck and the locomotive frame if they swing more. The rockers and swing bolster might not allow them to swing more than what they are designed to do for its minimum radius. The lead truck may hit on the cylinders or cylinder cocks.

Before time is spent turning flanges off of drivers, time and care should be spent looking at these other aspects to make sure other parts won't be damaged by going around tighter curves.

I want a large engine. Do you think I should build a *fill in the blank*?

For a beginner I recommend something easier than a large 4-8-4, 4-6-6-4, or 2-8-2, as with a larger engine, the parts are larger and at times a lot more complicated to machine. Generally speaking, the larger the engine the longer it takes to build and a small to medium size engine can take five or more years. I also recommend an engine that the builder can move easily and rerail easily as the engine won't always derail or need to be moved when there are other people around to help. While I enjoy running the 2-8-2, the 2-4-0 sees a lot more use as it is easier to get in, fire up, shut down and put away.

Appendix C: Common Myths

Valve gear is set for running in forwards so the engine is "square" and reverse is whatever because the valve gear can't be set square for forward and reverse. Also, engines don't use reverse much. (A related question: What does it mean that an engine is "square"?)

This myth is false. While it is possible for engines to be set up square in forward and really out of square in reverse, the gear is not designed that way. A combination of the gear being out of adjustment with slop in the valve gear can lead to this condition. When properly made and set, the valve gear will give proper admittance and exhausting of the steam for forward and reverse so forward and reverse are both square. When the valve gear is worn and the valve events are off, the valve gear will be set where it is typically set (hooked up, in the company notch) where the engine is run the most and is set for being even in that notch. Reverse (depending upon the valve gear) will be set as close as possible. Running hooked up will show inconsistencies in the valve gear much easier than running in the corner. Since running in the corner is very forgiving, the valve gear can be off a bit and still sound quite square when running in the corner. (Running square in the corner is easy. Running square hooked up takes more time and skill to set up.)

Engines will definitely need to run in reverse. Mainline engines generally run in forwards, but to do any kind of switching (with or without a train, to or from servicing facilities) they will need to run in reverse and may need to supply a good amount of power in reverse. Shortlines may not have any turning facilities so their engines could spend half their life running in reverse. Switch engines need to be able to move long heavy trains in forward and reverse.

Engines with two cylinders (like most engines built in the U.S. with one cylinder on each side [articulates included] or a two-cylinder Shay, Climax or Heisler) have the crank pins set nominally at 90 degrees. This way when one side is at front or rear dead center and applies no torque to the crank pin or wheels, the other cylinder is applying full torque to rotate the wheel. With the crank pins at 90 degrees and the valve gear set correctly, the engine will give four even exhausts per revolution of the drivers/drive shaft and is called square. When the valve gear is off and the exhausts are not even (not even spacing, one louder or softer than the others) the engine is out of square or lame.

Here are some videos of engines I have set or helped to set the valve gear on. The engines in all these videos are hooked up near center:

Walschaert valve gear—direct motion, 7.5" gauge: https://www.youtube.com/watch?v=tz0irFC7v70; https://www.youtube.com/watch?v=vvey7vCI3Fc; https://www.youtube.com/watch?v=KoyVVQgi0sI

Walschaert valve gear—direct motion, 15" gauge: https://www.youtube.com/watch?v=3OX6fd2WKJU

Running in reverse, 7.5" gauge: https://www.youtube.com/watch?v=HttXUpNxPoY; https://www.youtube.com/watch?v=GEjBEWt94vU

Walschaert valve gear—indirect motion, 24" gauge: https://www.youtube.com/watch?v=v-3J60sMl18

Stephenson valve gear, 15" gauge: https://www.youtube.com/watch?v=2-R_BSzxmgc

A good fireman never pops the safety and does not have dark or black smoke.

The safety valve should pop at least once a day to check it is working properly. There are times when the fireman anticipates the need for more steam and prepares the fire accordingly only to have something happen (the conductor waits for last-minute passengers, there is a delay on the track, etc.) where the demand is not needed. If there is not room to put water into the boiler, the safety can pop. Varying grades of coal can burn hotter, longer and/or faster than others. The first trip of the day is also a learning curve as to how that day's load of coal fires. It is also possible to have a mix where some fuel burns hotter than others.

As far as black or dark smoke, there are many factors to this. The type and grade of coal, the amount of draft, how much coal the fire needs, the type of anticipated work, etc. Some coal smokes a lot more than others so tossing on one scoop could create quite a bit of smoke where other coal may produce minimum smoke with multiple scoops of coal.

The more draft through the fire, the more coal can be added and burned without creating smoke due to more air coming through the fire. When the engine is working, the firebed needs to be thicker to prevent holes and cold spots in the fire and to produce the steam that is needed. When the engine stops, it is very possible for the stack to darken (even after having a clean stack when working for several minutes) as the flow of oxygen is less through the fire.

When anticipating a greater demand of steam when starting up, the fireman will start to prepare his fire to produce the steam when it is needed. Due to the light draft, the fire will produce smoke as coal is added.

Even if an engine burns oil, dark smoke can happen. If the oil is smoky and doesn't produce enough heat with a slight haze, the fireman will need to run with a smokier stack in order to get the heat that he needs from the fire.

The same philosophies hold true for live steam engines as well.

A good fireman never shakes the grates.

As the day goes on, the ash does build up in the firebox and the fire gets thicker. The ashier the coal, the thicker the fire gets. If an engine is really working hard, it will burn more coal and create more ash on the grates than if it is working easy. If there is ashy coal or the engine is working a long day, it would not be unreasonable to expect to shake the fire down at least once to clean it and get it to breathe better. The fireman should be firing per the conditions (work needed to be done, condition of fire and boiler, grade of coal, etc.) and sometimes the conditions are right to get a thick fire.

The same goes for the saying that a good fireman never needs to use the rake. If there is a clinker or the coal is not the greatest and fuses over or solidifies, then the rake may need to be used. If the coal is good and the fireman is on top of things, the chance of needing to use the rake is pretty low. However, with coals that are not as good, a rake may need to be used. When a small bank has been placed in the firebox (like for a dinner break) the rake will be used to break open and spread the bank to get the fire back.

On miniature engines, most of our grates don't shake but the grate bars are far enough apart to allow ample air to flow through the fire as well as allowing the ash to freely fall into the ashpan. With smaller fireboxes, the coal can fuse together or solidify much more easily than in full-size so the use of the rake is pretty much mandatory to break up the coal and get it to burn better as well as break up and remove any clinkers.

Slide valves wear in, spool valves wear out.

There has been a myth that slide valves are better than spool as with the steam pressure pushing down on the valve and into the seat, the valve will continuously "wear in" and "not leak or wear out" whereas with spool valves, the tension of the rings against the chamber will wear the bore big and will wear out the fit which will leak.

Slide valves and spool valves will slowly wear out over time. Proper lubrication will greatly slow down the wearing out process, but slide valves will wear out.

Slide valves and their seats will wear into convex shapes. There is a lot of pressure on slide valves due to the steam pushing them against their seats. As the valve is pushed forward, the front edge will wear into the seat more than the middle or rear of the valve. As the valve is pulled back, the back edge will wear into the seat more than the middle or front. Over time this means that the front edge of the valve and rear edge of the valve will wear the most where the middle won't wear nearly as much. This also means that the front edge of the seat and rear edge of the seat will wear more than the middle as well. Both the valve and seat will be worn into convex shapes and the valve will no longer seal on the seat.

This will happen over time with full-size and miniature engines. Proper lubrication will greatly slow down this process, but it will slowly happen.

Tall and sharp or square flanges are the best to keep the wheels on the track and not wear as fast.

The flanges should conform to IBLS standards. If the flanges are too tall, they will bottom out when running through switches and other places that have shallower flangeways. If they are too thick, the back-to-back will be too narrow if the gauge is set right and will bind on switches and guard rails. If the back-to-back is set right, the gauge will be too wide and will have interference with the track and on switches. The flanges should not have any sharp edges. The sharp edges will wear away the rail faster and the sharp edge will like to climb the rail and derail.

Water is a good lubricant for the cylinders, so a steam oil lubricator is not needed.

Water is a very poor lubricant. If you place water on one finger and rub it against another, your fingers will slide more easily than if they are dry. But if you do the same with oil, the fingers slide even more easily. The oil lubricates more than water, helping ease the friction between the fingers. The cylinders need steam oil (which is designed to withstand the temperature of the steam; motor oil will break down and not lubricate) to be properly lubricated. If they are not lubricated, the engine will be stiff and the rings and bores will wear and score prematurely.

Hooking up doesn't do anything in scale engines as the cylinders are too small.

Hooking up does reduce the steam admitted to the piston which reduces the amount of steam and water being used. The smaller the piston, the smaller the savings will be. Hooking up near the corner won't produce much difference in water usage, but hooking up near center, where the cut-off and port openings are drastically reduced from running in the corner, will produce results. The valve gear needs to be in good condition to do this, but people who operate 1:16 scale up to full-size see the benefits of hooking up. With my 4-6-0 I can run 18 trips around my track pulling five empty cars and have 1.5" of water left in the tender when running in the corner. When I hook her up to center, I can pull the same train 24 times around the track while using the same amount of water.

Keeping water high in the glass is safer than lower in the glass.

If the water is kept too high, the water can be easily carried over with the steam into the cylinders. The water can wash away the oil lubricating the cylinders and if enough water enters the cylinders, the cylinders can become hydrolocked and a lot of pressure can be generated as the piston nears the cylinder head. This pressure can damage or break the piston, cylinder block, cylinder head or bend a rod and the damage may not be readily apparent. Under normal conditions, a good water level is between one-third and two-thirds of a glass on level track. If there is water in the glass (and the glass is working properly) then the crownsheet is covered.

Appendix D: Inspections

Mechanical Inspections

It is a good idea to do a good visual inspection of the equipment on a periodic basis to look for defects such as loose nuts and bolts, loose/working pins, loose/working wheels on the axles, worn flanges, working joints, cracks, clearance on side bearings, spring rigging working well, etc. Following is a good basis on how to recognize some of these problems.

Loose nuts and bolts. Nuts or heads that have backed off away from the work, are only "finger tight," easily tighten with a wrench, when tapped with a hammer there is movement and/or does not sound tight and has more a dull "thud."

Loose/working pins. Pins that should be stationary can rotate or move in and out or side to side when tapped or when force is applied. Check for grease, oil or grime working out of the joint between the pin and the hole.

Loose/working wheels. Wheels moving in and out on axles, extra wobble in the rims, rods bind more as wheels wobble due to crank pin being at different angles, grease, oil or grime working out between the wheel center and axle (key is not meant to do the driving).

Worn flanges and wheels. Sharp, thin and/or tall flanges. No fillet between tread and flange. No taper on tread, tread hollowed out. Derailments can chip cast iron flanges. Sharp flanges increase the risk of derailments.

Working frame joints. Grease, oil or grime working out of joints in frame. Frame bolts may be tight or may be loose.

Axles out of tram. Crank pins bind when pins pass over dead center. Uneven flange wear, too much flange wear compared to time running and running conditions, axles are not perpendicular to track when on straight track, centerline measurements are not the same between the right wheels and the left wheels of the same axle sets.

Cracks. Cracks may be visible to the naked eye and can appear as new lines that grow. Grease, oil and grime can also work out of them. The cracks will form perpendicular to the force acting on the piece. The force on a main/side rod is horizontal so the crack will be vertical. On a pedestal the crack will be horizontal. Any sharp corner (between pedestals and the frame, corners on truck frames are some examples of where sharp corners are) is a stress riser and cracks can radiate out from the corner. Cracks can radiate out from where axles and crank pins are pressed into the wheel. Cracks can begin at the sharp corners where the keys are in the driver and radiate out from the corners.

Side bearing clearance. In a perfect world, the weight of the car would rest on the center of the bolster and there would be equal clearance between the car frame and both side bearings when sitting on flat, level track. When looking at the side bearings, if one side bearing is touching, the other side bearing should have clearance.

Spring rigging. When leaf springs are used, are the ends of the leaf springs at the same height? Are any leafs broken or dislodged? Any cracks in the equalizers and spring hangers? The engine should be able to rock side to side and the spring hangers should not hang up on anything.

Driving and journal boxes. The driving boxes and journal boxes should move freely in the frame. Prying up on the wheels will show how easily the boxes move. The springs should flex as the wheels are picked up. With a truck, one wheel should be able to be picked up a fair distance while the other three remain on the rails. However, with trucks, side frame should not be able to move forward and backwards while the other truck remains still. The bolster should hold the side frames solidly in tram.

Lubrication. With steam engines, the piston and valve rods and the bores should have a thin film of oil. Any bearing surfaces should also have a thin layer of film and have no sharp edges on the bearings or bearing surfaces which can wipe oil away. If bearings get hotter than usual or typically run hot, there is a problem with the lubrication or alignment.

Side rod issues. If rods bind over dead centers, there is most likely a problem with tram or the throw of the crank pins are not the same. If rods bind over top and bottom quarter, there is most likely a problem with the quartering.

Pounding around the cylinders. Once per revolution is most likely a piston that is too far forward or too far to the rear in the bore and is hitting on a cylinder head. A couple times per revolution could be a loose piston. Pounding can also be a driving box that has too much clearance between the pedestals.

Boiler Inspections

In full-size, periodic inspections are done in addition to daily inspections to look for leaks at seams, cracks in sheets, leaks around staybolts and flues and the overall condition of the boiler. With boilers, failures are preceded by warning signs that something is wrong. The warning signs should be heeded immediately and repairs done. While some of these items apply directly to boilers, others can also apply to other pressure vessels such as air tanks. Here is a basic overview of what to look for.

Sight glass inspection. The sight glass should be blown down in a manner which will confirm that the water and steam passageways are free flowing and that the steam and water valves to the glass seal. (1) Open drain. (2) Close water valve, steam blows out drain. (3) Close steam valve. Nothing should exit drain. (4) Open water valve, glass should fill with water. (5) Open steam valve. (6) Close drain, water should return quickly to previous level. This should be done daily. On a periodic inspection (FRA requires this every 31 service days, a day counts as a service day if there is pressure in the boiler and fire in the firebox), the piping and valves to the glass should be cleaned of scale.

Gauge cocks. The gauge cocks should be used to make sure that they function properly. If the gauge cock is exposed to water, water will spurt out the gauge cock and some of the water will boil into steam. If the gauge cock is exposed to water and steam, a mixture of water and steam will spurt out. If the gauge cock is exposed to steam, steam will spurt out, but there will also be some water as the steam condenses. On a periodic inspection (31 service day inspection for FRA) the gauge cocks should be taken apart and any scale should be cleaned.

Firebox inspection. Done each day when the fire is calm or dumped so the heat from the fire won't evaporate water leaking out. Look for water leaking (and listen for the telltale hiss) around tubes in the tube sheet, around the staybolts and crownstays, signs of cracking of the side sheets around the bolts, leaks in the sheets or bulges in the sheets.

Smokebox inspection. Look for leaks around the tubes, stays, seams and steam piping. Also look for holes in the smokebox which will allow air to enter and cut down the draft through the fire.

Boiler washes. Boiler washes should be done on a periodic basis (FRA requires one every 31 service days) where the plugs are removed from the backhead, barrel and mud ring. Wands, hoses or tubing is then inserted through these washout plug holes (starting at the top) to wash scale and boiler mud off of the crownsheet, tube pack, side sheets and barrel and wash it down to the mud ring where it is washed and racked out. There might be 10–25 washout plugs on the boiler in order to gain access to the sheets, stays, tube pack and the barrel in addition to the four washout plugs along the mud ring. On some miniature engines, the steam dome cover can be removed for washing and inspecting the tube pack and inside of the boiler.

Draining. The pressure vessel should be drained of water when being stored to prevent water collecting in the bottom and rotting out the bottom of the pressure vessel. Low lying piping should also have drains to drain out water.

Hydro test. One test that is used to test a pressure vessel is the hydro test, where the vessel is pressurized with water (not air or steam, as with water, if the vessel fails, water will squirt out. Water doesn't compress but air and steam do; a failure can cause an explosion) typically to a pressure 1.25–1.5 times the working pressure. The vessel is then gone over looking for leaks, bulges, weeping, etc. Just because a pressure vessel passes a hydro test does not mean it is totally safe. The vessel might have enough strength to not fail, but its strength might not be up to par to what it was designed to be and should be for safe operation. Hydro tests should not be the only type of inspection/test used to determine if the boiler is safe and good to use.

Internal inspection. While doing an internal inspection, the tube pack and boiler sheets should be inspected for scale build-up. Scale will hold moisture against the steel and promote rotting as well as not allowing the heat from the fire to pass through the metal into the water as easily. This could cause localized areas of the steel to overheat.

Inspect sheets for pitting, blisters, burning (blackened steel), thinning, grooving, cracks, necking down of staybolts or crownstays, erosion of metal (where dissimilar metals are used, one metal will erode and be sacrificed in a process called electrolysis. Zinc anodes are used when copper tubes are used in a steel boiler for this reason) and any signs of bulging or metal pulling away from seams or bolts. Pitting that is dark in color is typically the result of oxygen in the water eating away at the metal. It also leaves hollows where water can collect when pressure vessel is drained, which will increase pitting

The bore scopes that are available to the public today are a lot better than the ones that were available 15 or 20 years ago. Having a bore scope is a good way to be able to see inside the boiler in areas that are not visible otherwise.

UT survey. It is also a good idea to map out the metal of the pressure vessel and take

UT measurements of the sheets and document them. Wastage of the metal thickness can be tracked, the print thickness on the blue prints can be compared to the material that was actually used, thin spots can be found and repaired before they become a problem, and if the pressure vessel is old, the thickness of the sheets should be used (along with the worn diameters of stays) to calculate if the pressure vessel is still safe to use.</APP>

Why Internal Boiler Inspections Are Important: Some Lessons Learned from the Medina, Ohio, Accident (July 29, 2001)

Passing a hydro test does not mean that a boiler is safe to fire or pressurize. Passing a hydro test means that the boiler did not fail at 1.25 or 1.5 times the operating pressure, but other than that, it does not signify what condition the boiler is in. Sheets may be cracked, corroded, thin or bowed; staybolts, crownstays and braces may be thin, cracked, broken and the sheets which they are threaded into may be so thin that only a couple of threads are holding. Seams may be cracked, thin or leaking. Internal inspections are meant to find these problems before something bad happens to the boiler.

The traction engine boiler in the Medina, Ohio, accident on July 29, 2001, had a thin spot on the crownsheet. It was worn down to a thickness of 0.087" from the original 0.375" thickness. The stays should have had 4.5 thread engagement with the original sheet thickness, but was down to 2.5 thread engagement on average. At least five stays were welded around the thread due to the loss of metal in the crownsheet. The stays were corroded to 0.6"–0.7" diameter, which resulted in a 64 percent reduction of the cross sectional area of the bolts. The thin sheet and corroded bolts greatly weakened the boiler. The report can be found here: http://www.farmcollector.com/steam-traction/final-report-tragedy-medina-county-fairgrounds.

It is noted in the report that at least one inspection plug/port had not been removed in a long time, showing that a proper internal inspection could not have been conducted in the recent past. Per calculations based off the condition of the boiler, the boiler should have had an operating pressure between 40–47 psi (safety was "set" at 125 psi, but it did not pop when tested).

This report also gives an example of how the boiler could have failed even if the crownsheet had been covered with water (fusible plug not melting); the thin metal bagged between the blots which allowed a low spot for scale to collect and insulate the sheet from the water. The sheet then is not cooled by the water, overheats slightly and bags slightly and the process is repeated until mechanical failure occurs. The report stated that this does not support the Boiling Liquid Expanding Vapor Explosion. Internal inspections and using a UT machine are meant to catch large or major defects such as these.

I have read in various reports and have met and talked with friends of the former owner and they said, "If he only knew about that he would not have run the engine."

Even small boilers can corrode away and parts can fail. A boiler that has failed due to thin sheets, corroded bolts, etc., is an example of someone not taking care of the boiler. Wastage of sheets doesn't happen overnight.

Appendix E: Tractive Effort and What an Engine Can Pull

Tractive Effort versus Grades and Curves

It seems backwards that the 2-4-0 has pulled 17 cars without much problem while the 2-8-2 struggled with 31 cars. The difference is the track conditions. The 17 cars with the 2-4-0 was on level track and the 31 cars with the 2-8-2 was on an S curve with a 2 percent+ grade.

The max tonnage a steamer is calculated to pull is the tractive effort divided by resistance. On straight, level track the resistance is four pounds per ton.

Note: These calculations are based off of the power that the engine can produce. Whether or not the wheels have enough weight, and thus enough adhesion, to transmit that power to the rails is another thing.

The 2-4-0 has a calculated 140 pounds tractive effort, so on straight, level track she can pull (140/4) = 35 tons or 70,000 pounds. This includes the weight of the engine and tender.

Each percent grade adds an additional 20 pounds of resistance. So on a straight 1 percent grade, she can pull (140/24) = 5.83 tons or 11,667 pounds.

Each degree of curvature adds an additional 0.8 pounds of resistance.

My track has a 3 percent grade with a 52' radius (148-degree) curve. The total resistance per ton is $4 + (20 \times 3) + (0.8 \times 148) = 182.4$ pounds/ton.

This means on my track the 2-4-0 can pull (140/182.4) = 0.76 tons or 1,520 pounds. The max I have pulled was around 1,500–1,530 pounds and I put sand on the track to help with adhesion, but that also increases resistance. These calculations are pretty accurate for the miniature railroads as well.

Formulas

Tractive Effort

Tractive effort (at the drawbar in pounds) is calculated using the area of the piston (in square inches), stroke (inches), boiler pressure (psi), number of power strokes per revolution of the drivers and the diameter of the drivers (inches). Tractive effort is the theoretical maximum power that the cylinders can produce.

Tractive Effort (TE) = (area of piston)(stroke)(# of power strokes per rev of driver)(boiler pressure psi)(.85)/(driver circumference).

This may also be expressed as: TE = (π × piston radius squared)(stroke)(# of power strokes per rev of driver)(boiler pressure psi)(.85)/(π × driver diameter).

This may be expressed after factoring out π: TE = (piston radius squared)(stroke)(# of power strokes per rev of driver)(psi)(.85)/(driver diameter).

Resistance and How Much an Engine Can Pull

The amount a steam engine can pull (in tons and including the weight of the engine/tender) is tractive effort divided by resistance.

Resistance on straight, level track is 4 pounds per ton.

Resistance on grade = (20 pounds/percent grade)(percent grade) + 4 pounds/ton; a 2 percent grade is 0.02 for the calculation, or an additional 40 pounds of resistance from level.

Resistance for curvature adds an additional 0.8 pounds per degree of curvature.

Degree curvature: How many degrees a track curves when spanned by a 100' chord; a 180 degree (50' radius) curve is the tightest that can be calculated this way. Degree curvature = 2arcsin(50/radius).

Total resistance (pounds per ton) = 4 + 20(percent grade) + .8(degree curvature).

My track has a 52' radius curve (148 degrees) with a 3 percent grade. On that curve with that grade it is 4 + 118 + 60 =182 pounds/ton resistance.

Another note: This is also based off of a continuous grade. A short (like 100' or shorter), steep grade is not as difficult as a steep continuous grade (several hundred to thousands of feet). With a short, steep grade or short curve, it is possible that not all of the train is on the grade so the resistance will not be continuous. On a long grade or curve, the resistance will be continuous.

With a short grade, the hill can be entered with speed and the speed will drop, but not stall before exiting the steepest grade or tightest curve so the speed exiting will be less, but speed can't be maintained.

The 4-6-0 is rated around 1,900+ pounds on my track, but I have pulled over 2,050 pounds up the grade and around the curve. She slowed down in the curve and started regaining speed after the curve but still on the 3 percent. She would not start that train on the curve and grade nor could she pull that train up too much longer of a curve with that grade as she would stall. She did not have enough power to maintain speed with the 2,050-pound train.

Tractive Effort versus Factor of Adhesion

The pulling power of steam engines is rated in pounds of tractive effort. This is a calculation using area of the piston, stroke, number of power strokes per revolution of the driver, boiler pressure and driver circumference. The tractive effort is the amount of power that the steam engine is capable of producing.

The factor of adhesion is the ratio between the weight on the drivers and the tractive effort: weight on drivers / tractive effort. The full-size aimed for a factor of adhesion of around 4, meaning that for every 1 pound of tractive effort there would be 4 pounds on the drivers. The lower the factor of adhesion the slipperier the engine would be.

Many scale engines have tractive efforts that are a lot larger than scale. CB&Q 2-8-2 #4960 has around 58,090 pounds of tractive effort. This would scale to 137.69 pounds of tractive force for a 1.6" scale 2-8-2. A 1.6" scale is 1:7.5. To find the true scale tractive effort if all linear dimensions are perfectly scaled it is $58{,}090/(7.5^3)$.

A Railroad Supply 1.6" scale 2-8-2 has 286 pounds of tractive effort. Boiler pressures are rarely scaled down to the true scale. To compensate for the higher boiler pressure, the piston diameter is made smaller to reduce the tractive effort.

However, with having such a high tractive effort, since there is about 700 pounds on the drivers, the factor of adhesion is around 2.44, meaning that she can be a slippery engine if all the power is applied to the drivers.

The 2-8-2 is around 900 pounds which would scale up to around 190 tons for the engine. That is pretty reasonable. So the reason why live steamers are slippery is not due to the lack of weight but due to the increased tractive effort.

While an engine can produce a lot of tractive effort that does not mean it can pull huge trains. There needs to be enough weight on the drivers to allow all the power generated by the cylinders to be applied to the rails. Adding more weight to the drivers or adding sand to the rails does not increase the tractive effort but allows the engine to apply all of its power to the rails so more power can be applied to the rails (so an engine can pull more) without slipping the drivers.

Tractive Effort versus Driver Diameter

It has been said that engines that were used for yard service and freight had smaller drivers to pull more whereas passenger engines had larger drivers for speed. The larger the driver, the faster the engine can run. But if the cylinder proportions stay the same, the larger the driver the less an engine can pull.

The formula for tractive effort is (area of piston)(stroke)(.85)(boiler psi)(# of power strokes per revolution)/(driver circumference).

π can be taken out of the formula and simplified to: (radius of piston squared)(stroke)(.85)(boiler psi)(# of power strokes per revolution)/(driver diameter).

Let's assume an engine has 2" diameter pistons, 3.25" stroke and operates at 125 psi and has the normal 4 strokes per revolution.

If the drivers are 6", this gives a tractive effort of 230 pounds (can pull 10,464 pounds up a straight 2 percent grade).

If the drivers are 8", this gives a tractive effort of 172 pounds (can pull 7,848 pounds up a straight 2 percent grade).

A Little Engines 0-4-0 (2.375" × 3" cylinders, 7" driver) has 256 pounds of tractive effort, where a Little Engines 4-4-2 (2.5" × 3.5" cylinders, 10" driver) has 232 pounds of tractive effort (when they are at 125 psi). Based on that, the LE 0-4-0 can pull more than a LE 4-4-2, but since the 4-4-2 has a lead truck, the 4-4-2 will have better stability and due to the larger boiler, will have better steaming capacity. If the engine doesn't have enough steaming capacity to keep up with the demand placed upon it, the boiler pressure will fall and so will the tractive effort. A small engine such as a switcher (0-4-0, 0-6-0) and early style 4-4-0s have small boilers, so while they may have ample tractive effort, on long steam grades they might not be able to produce all of the tractive effort if the boilers can't produce steam as fast as it is being used.

Index

www.ingramcontent.com/pod-product-compliance
Ingram Content Group UK Ltd.
Pitfield, Milton Keynes, MK11 3LW, UK
UKHW051929100926
8556IPUK00022B/1456

9 781476 672816